Reaction Dynamics

M. Brouard

University Lecturer in Physical Chemistry and Tutor
and Fellow of Jesus College, Oxford

Series sponsor: **ZENECA**

ZENECA is a major international company active in four main areas of business: Pharmaceuticals, Agrochemicals and Seeds, Specialty Chemicals, and Biological Products.

ZENECA's skill and innovative ideas in organic chemistry and bioscience create products and services which improve the world's health, nutrition, environment, and quality of life. ZENECA is committed to the support of education in chemistry and chemical engineering.

OXFORD NEW YORK TOKYO
OXFORD UNIVERSITY PRESS
1998

Oxford University Press, Great Clarendon Street, Oxford OX2 6DP

Oxford New York
Athens Auckland Bangkok Bogota Bombay
Buenos Aires Calcutta Cape Town Dar es Salaam
Delhi Florence Hong Kong Istanbul Karachi
Kuala Lumpur Madras Madrid Melbourne
Mexico City Nairobi Paris Singapore
Taipei Tokyo Toronto Warsaw

and associated companies in
Berlin Ibadan

Oxford is a trade mark of Oxford University Press

Published in the United States
by Oxford University Press Inc., New York

A catalogue record for this book is available from the British Library

Library of Congress Cataloging in Publication Data
(Data applied for)

ISBN 0 19 855907 0

Typeset by EXPO Holdings, Malaysia
Printed in Great Britain by
Bath Press, Bath

Series Editor's Foreword

Oxford Chemistry Primers are designed to provide clear and concise introductions to a wide range of topics that may be encountered by chemistry students as they progress from the freshman stage through to graduation. The Physical Chemistry series will contain books easily recognized as relating to established fundamental core material that all chemists will need to know, as well as books reflecting new directions and research trends in the subject, thereby anticipating (and perhaps encouraging) the evolution of modern undergraduate courses.

In this Physical Chemistry Primer, Mark Brouard has produced an authoritative and clearly written account of *Reaction Dynamics* detailing fundamental aspects of chemical reactions occurring in the gas phase. This Primer will stimulate all students of chemistry and their mentors.

Richard G. Compton
Physical and Theoretical Chemistry Laboratory,
University of Oxford

Preface

The study of the dynamics of elementary gas phase reactions addresses fundamental questions about how atoms and molecules interact and undergo chemical change. The aims of this Primer are to introduce reaction dynamics at an amenable second or third year undergraduate level, and to provide a unified treatment of the dynamics and kinetics of elementary chemical reactions. Many of the key concepts are illustrated by recent experimental results, and an extensive list of references to the original scientific literature is provided.

I would particularly like to thank Javier Aoiz and David Manolopoulos, both for supplying data for figures, and for critically reading the Primer. I am also indebted to Simon Gatenby for proof-reading the manuscript, and for helping with the drawing of numerous figures.

Oxford M. B.
October 1997

Contents

1 Introduction 1
- 1.1 Elementary reactions versus complex reactions 1
- 1.2 The elementary processes 1
- 1.3 Reaction kinetics and dynamics 2
- 1.4 From cross-sections to rate coefficients 5

2 Potential energy surfaces 8
- 2.1 Calculation of potential energy surfaces 8
- 2.2 Types of potential energy surface 10
- 2.3 Experimental probes of potential energy surfaces 14
- 2.4 Motion over the surface 15

3 The differential cross-section 18
- 3.1 Elastic scattering 18
- 3.2 Reactive scattering 22
- 3.3 Case studies 28
- 3.4 Stereochemistry 31

4 State-specific cross-sections 33
- 4.1 Experimental considerations 33
- 4.2 Models of energy utilization and disposal 34
- 4.3 Kinematic constraints 39
- 4.4 Case studies 44

5 Microcanonical rate coefficients 46
- 5.1 The cumulative reaction probability 46
- 5.2 Transition state theory (TST) and $N^{\ddagger}(\epsilon)$ 49
- 5.3 $k(\epsilon)$ for unimolecular reactions 52
- 5.4 The measurement of $k(\epsilon)$ 58

6 Thermal rate coefficients 60
- 6.1 Canonical transition state theory (CTST) 60
- 6.2 Thermally activated unimolecular reactions 66

Appendices 72
- A.1 Classical elastic scattering of two atoms 72
- A.2 Quantum mechanical reaction cross-section 74
- A.3 Cumulative reaction probability 74

Bibliography 76
- Background reading 76
- References 77

Index 80

1 Introduction

1.1 Elementary reactions versus complex reactions

A central goal of chemistry is to understand molecular change. How do molecules interact with each other and evolve into new chemical species? Why are some reactions slow and others fast, and why do the chemical reaction rates depend on experimental conditions such as temperature and pressure? These questions can be answered most comprehensively for chemical reactions occurring in the gas phase, where complex interactions of the reacting species with the environment (such as a solvent in the liquid phase) are minimized. However, even in the gas phase, the chemistry which takes place may be highly complex, involving numerous individual reactive (and non-reactive) processes. This complexity is typified by the gas phase chemistries involved in combustion and in the atmosphere.

The focus of this book is elementary gas phase reactions. If it were possible to measure or, ideally, to predict *a priori* the rates and products of these fundamental processes, then the more complex gas phase chemistry, involving many hundreds of elementary steps, might be modelled. Indeed, without such detailed knowledge of elementary reactions it would be impossible to provide a quantitative understanding of complex gas phase chemical processes.

1.2 The elementary processes

Most elementary reactions can be categorized as either unimolecular or bimolecular, and it is these processes which will be dealt with in detail. Unimolecular reactions can be represented by the equation

$$\mathrm{A} \longrightarrow \mathrm{A}^* \longrightarrow \text{products}\,,$$

which emphasizes that a stable molecule, A, can only undergo reaction if it acquires sufficient energy to overcome an energy barrier to reaction. The reactant may gain this energy in a variety of ways. One way is via a bimolecular collision with an unreactive molecule, M (an *inelastic*, energy-transferring collisional process),

$$\mathrm{A} + \mathrm{M} \longrightarrow \mathrm{A}^* + \mathrm{M}\,,$$

which may be termed thermal activation. Alternatively, A may be energized by the absorption of light; the overall reactive process is then called photodissociation. Although examples of both thermally activated and photon-induced unimolecular reactions will be presented, more comprehensive accounts of photochemistry and photodissociation dynamics can be found elsewhere (see Background reading).

A final class of reactions is *association reactions*,

$$\mathrm{A} + \mathrm{B} \rightarrow \mathrm{AB}^* \xrightarrow{\mathrm{M}} \mathrm{AB}.$$

These reactions may be regarded formally as the reverse of unimolecular reactions. Under certain conditions, the species AB^* may have sufficient energy to undergo further reaction to form new products C and D, and the overall process is then called *chemical activation*. Association reactions and chemical activation can be modelled using the same theories as those developed to rationalize unimolecular reactions, and will not be discussed in detail.

1.3 Reaction kinetics and dynamics

Thermal rate coefficients

The rate of a bimolecular reaction between species A and B is proportional to the reactant concentrations, [A] and [B], the proportionality constant being the rate coefficient, $k(T)$, as defined by the rate law,

$$-\frac{\mathrm{d}[\mathrm{A}]}{\mathrm{d}t} = k(T)\,[\mathrm{A}]\,[\mathrm{B}]\,.$$

The magnitude of the rate coefficient depends on a variety of factors under experimental control, most notably the temperature. Observations of the variation of rate coefficients with temperature led Arrhenius to propose the following empirical equation for $k(T)$:

An historical account of the development of this and other equations can be found in Laidler's *Chemical kinetics*.

$$k(T) = A\mathrm{e}^{-E_a/RT}\,. \tag{1.1}$$

In general, the activation energy for the reaction, E_a, is defined by the equation

$$\frac{\mathrm{d}\ln k(T)}{\mathrm{d}T} = \frac{E_a}{RT^2}\,. \tag{1.2}$$

Simple collision theory

Derivations of simple collision theory can be found in many of the general texts referred to in the Background reading. A more detailed derivation will be given in Chapter 4.

One of the first, partially successful attempts to rationalize the form of the Arrhenius expression, *simple collision theory*, recognized that reactions take place via collisions. A simplified derivation of the collision theory expression for the rate coefficient of a bimolecular reaction proceeds by equating the reaction rate with the rate of collisions, times the fraction of collisions which have sufficient *kinetic* energy to surmount the reaction barrier, ϵ_0 (see Chapter 2). The resulting expression may be written

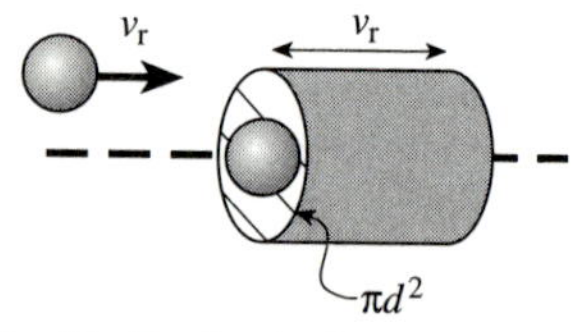

Fig. 1.1 The hard sphere collision cross-section, πd^2. The collision volume, $v_r\sigma_c$, also shown, is that swept out in unit time.

$$k(T) = Z^0_{\mathrm{AB}}\,\mathrm{e}^{-\epsilon_0/k_BT}\,, \tag{1.3}$$

where Z^0_{AB}, the collision number or the collision frequency factor (the collision frequency divided by the reactant number densities), is given by

$$Z^0_{\mathrm{AB}} = \left(\frac{8k_BT}{\pi\mu}\right)^{1/2}\sigma_c\,.$$

It is the *relative* velocity of the reactants (A and B) which determines the collision kinetic energy, $\epsilon_t = \frac{1}{2}\mu v_r^2$, where $\mu = m_A m_B/m_{AB}$, with $m_{AB} = m_A + m_B$ (see Section 2.4).

The first term in the above equation is the average *relative* speed of the reactants, $\langle v_r\rangle$, while the second term is the collision cross-section. The latter, which has dimensions of area, is the area within which reactants A and B must approach one another for collision to take place, as illustrated in Fig. 1.1. In simple collision theory, this is estimated assuming the reactants are hard spheres and is given by πd^2, where the collision diameter, d, is the sum of the radii of the two reactants, $d = r_A + r_B$.

Table 1.1 Comparison between experimental and simple collision theory *A*-factors. *P* is the steric factor referred to in the text, and is simply the ratio of the experimental and calculated *A*-factors. Data from Pilling and Seakins' *Reaction kinetics*

Reaction	T/K	E_a/ kJ mol^{-1}	$10^{-11} A_{expt}$/ dm^3 mol^{-1}s^{-1}	$10^{-11} A_{sct}$/ dm^3 mol^{-1}s^{-1}	P
$K + Br_2 \rightarrow KBr + Br$	600	0	10.0	2.1	4.8
$CH_3 + CH_3 \rightarrow C_2H_6$	300	0	0.24	1.1	0.22
$2\,NOCl \rightarrow 2\,NO + Cl_2$	470	102	0.094	0.59	0.16
$H_2 + C_2H_4 \rightarrow C_2H_6$	800	180	1.2×10^{-5}	7.3	1.7×10^{-6}

Although the form of the rate coefficient obtained from simple collision theory is gratifyingly similar to that given by the Arrhenius expression (suggesting that the model is at least on the right track), detailed comparison of the rate coefficients predicted by eqn 1.3 with experimental data reveals major discrepancies, primarily associated with the predicted *A*-factor, which may differ from the experimentally derived values by many orders of magnitude. Some representative data are given in Table 1.1, in which the simple collision theory estimates have been made assuming the barrier height is equal to the experimentally determined activation energy, and the collision radii have been obtained from viscosity data. Also introduced in this table is the *steric factor*, *P*, which is defined empirically as the experimental divided by the calculated Arrhenius *A*-factors.

In reality, molecules cannot be regarded as hard spheres, and the disagreement between experiment and the simplest theory may not appear surprising. However, the discrepancies are too large to be accounted for simply by improper calculation of the *collision* cross-section, as is clear from the large deviations of the steric factor from unity. The underlying problem is with the definition of a reactive collision as a collision with sufficient *kinetic energy* to surmount the reaction barrier. Reactions, as opposed to non-reactive collisions, may require reactant molecules to approach in preferred configurations or orientations to enable the electrons to rearrange and thus allow bonds to be broken and new bonds formed. As the complexity of the reactants increases, the chances of them approaching in the correct orientation to react become diminished, and reactive collisions become an ever-decreasing subset of all collision encounters, as reflected by the decreasing magnitude of the steric factor.

The difference between the reaction and collision cross-sections could be accommodated in the crudest fashion by replacing the latter, σ_c appearing in eqn 1.3, by $\sigma_r = P\sigma_c$, where *P*, the steric factor, represents the *average* reaction probability on collision.

Reaction cross-sections

To provide some insight into the significance of the reaction cross-section, consider the reactive collision encounter in more detail. Imagine a collision between two reactant molecules, A and B, as illustrated in Fig. 1.2. The reactants approach with relative velocity $\mathbf{v}_r$, which may be oriented so that the reactants collide head-on (i.e. along the line connecting the centre-of-masses of the two reactants), or with a 'glancing blow' collision, as illustrated. The difference between these two encounters is quantified by the *impact parameter*, *b*, of the collision, which is defined as the distance of closest approach of the reactants in the absence of any interactions between them.

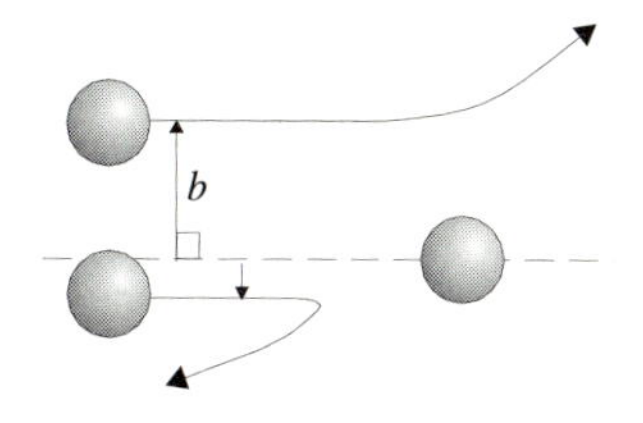

Fig. 1.2 Definition of the impact parameter.

More fundamentally, the impact parameter defines the classical *orbital angular momentum* of the reactants: see Chapter 3.

The definitions presented here for the reaction cross-section are derived from classical mechanics; see Chapter 3.

With this definition, a head-on collision occurs when $b = 0$, and a 'glancing blow' collision occurs when $b > 0$.

In general, an experimenter attempting to measure the reaction rate coefficient or cross-section has no control over the magnitude of the impact parameter at which any particular collision takes place. The reaction cross-section (the reactive target area of the reactants) must represent an average over collisions with different impact parameters. Furthermore, reactive collisions form only a subset of all collisions: they are those collisions which lead to reaction as opposed to energy transfer. Therefore, as well as averaging collisions with different b, the reaction cross-section must also account for the probability of reaction on collision at a specified impact parameter, $P(b)$. The correct averaging leads to the following expression for σ_r

$$\sigma_r = \int_0^{b_{max}} P(b)\, 2\pi b\, db, \tag{1.4}$$

It may be noticed that even the opacity function, $P(b)$, is an average reaction probability over reactant encounters with different relative orientations.

where $P(b)$, the probability of reaction on collision at a specified impact parameter, is known as the *opacity function*. The upper limit of the integral, b_{max}, implies that there is a finite range of impact parameters which can lead to reaction; beyond b_{max}, the collisions are so glancing that the probability of reaction is vanishingly small. The use of $2\pi b\, db$ to represent the 'volume element' for integration reflects the fact that the experimenter has no control over the azimuthal angle, ϕ, at which the collision occurs. The reaction cross-section, therefore, may be viewed as a 'dartboard' average of the reaction probability, $P(b)$, over a finite range of impact parameters. The ϕ azimuthal angle and the origin of the $2\pi b\, db$ factor are illustrated in Fig. 1.3.

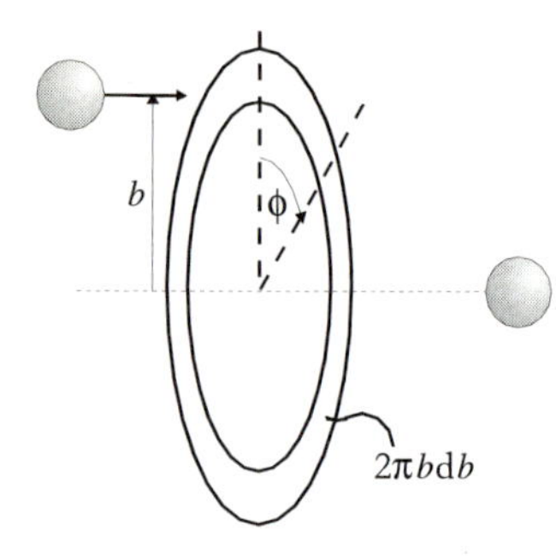

Fig. 1.3 The dartboard averaging over impact parameter.

To demonstrate that eqn 1.4 provides a reasonable definition of the reaction cross-section, take the simplest choice of opacity function, $P(b)$,

$$P(b) = 1 \qquad b \leq b_{max},$$

and zero otherwise. The resulting cross-section is given by

$$\sigma_r = \int_0^{b_{max}} 2\pi b\, db = \pi b_{max}^2,$$

which has the correct dimensions of area per molecule, already discussed. This simple expression allows estimation of the maximum impact parameter for reaction from experimentally determined reaction cross-sections.

State-to-state reaction cross-sections

So far, no mention has been made of the reactant molecules undergoing reactive collisions. It seems reasonable to suppose that the probability of a reaction on collision, and hence the reaction cross-section, depends on the kinetic energy of the reactants, and also on their internal rotational and vibrational state. For example, if a particular reactant bond is to be broken in the course of reaction, initial vibrational energy in that reactant bond might preferentially enhance its cleavage. If that were so, then the reaction probability and cross-section would depend not just on the total energy of the reactant molecules, but also on the degrees of freedom (reactant translation, rotation, or vibration) in which this energy is initially stored.

For now let us simply acknowledge the possibility of such a dependence, and introduce the notation by which it may be characterized. Consider the model triatomic system, undergoing an atom plus diatom bimolecular reaction at a specified collision kinetic energy,

$$A + BC(v, j) \longrightarrow AB(v', j') + C,$$

where the reactant and product vibrational and rotational states are labelled v, j and v', j', respectively. *State-to-state* reaction cross-sections, $\sigma_{if}(v_r)$, are introduced to characterize the variation in the reaction cross-sections with vibrational (v and v') and rotational (j and j') states of the reactant and product molecules, and their dependence on collision energy, $\epsilon_t = \frac{1}{2}\mu v_r^2$ (or relative velocity, v_r). For simplicity of notation, the internal states of the reactants and products are labelled 'i' and 'f' respectively. State-to-state reaction cross-sections may again be interpreted as reactive target areas, although on this occasion the reactant molecules approach in well-defined initial quantum states, and the products depart in specific final states.

The energy and internal state dependence of reaction cross-sections and rate coefficients is discussed in Chapter 4. It will be seen that reaction cross-sections also depend on the internal states of the reaction products, and we allow for such a dependence in what follows.

Differential cross-sections

When two molecules collide and react, the products are scattered in a variety of directions (*scattering angles*), relative to the direction of reactant approach. This directional property of chemical reactions is illustrated for the simple case of atom–atom collisions by Fig. 1.4. The quantity which characterizes the angular dependence of the reaction cross-section is known as the *differential cross-section* (with dimensions of area per solid angle, $d\omega = \sin\theta d\theta\, d\phi$), $d\sigma_r/d\omega$. The 'integral' cross-sections of the preceding sections are obtained from differential cross-sections by integrating over scattering angles (θ and ϕ),

$$\sigma_r = \int_0^{2\pi} \int_0^{\pi} \frac{d\sigma_r}{d\omega} \sin\theta d\theta\, d\phi. \tag{1.5}$$

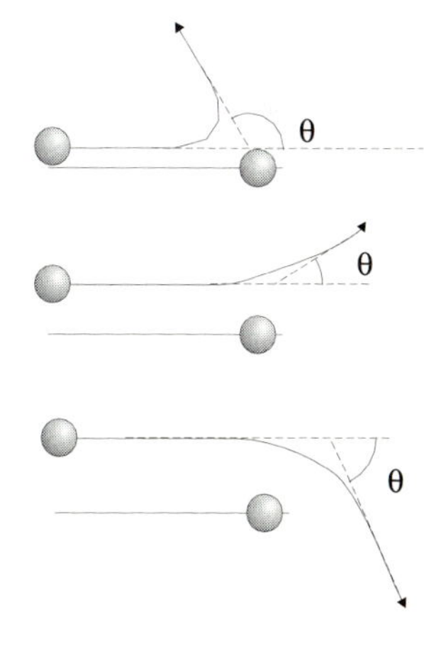

Fig. 1.4 Collisions lead to scattering with a distribution of scattering angles.

Once again, the quantity $d\sigma_r/d\omega \sin\theta d\theta\, d\phi$ can be regarded as the reactive target area, but this time for reactions which lead to products scattered in the angular range $\theta \to \theta + d\theta$ and $\phi \to \phi + d\phi$.

The importance of the differential cross-section lies in the fact that it is sensitive (albeit indirectly) to properties which are not otherwise amenable to experimental study, such as the opacity function (see Chapter 3). As a consequence, it is one of the most detailed *measurable* properties of a chemical reaction that can be compared with theoretical prediction.

1.4 From cross-sections to rate coefficients

Thermal averaging

The term *thermal* rate coefficient implies that the reactant molecules are in thermal equilibrium at a specific temperature. At thermal equilibrium, molecules possess distributions of kinetic, rotational, and vibrational energies and, thus, collisions too must necessarily involve species endowed with distributions of relative speeds and internal quantum states. The thermal rate coefficient must reflect the average outcome of these many different collision encounters.

Thermal equilibrium may not always be maintained, particularly in the internal vibrational modes of the reactants at low pressures. If it is not, because the rate of energy-transferring collisions (which are responsible for restoring the system to equilibrium) is small relative to the reaction rate, alternative averaging procedures are required to determine the reaction rate.

Statistical mechanics provides a precise definition of thermal equilibrium. The relative translational motion of the reactants at thermal equilibrium is characterized by the Maxwell–Boltzmann distribution of relative speeds,

$$f(v_r)\,dv_r = \left(\frac{\mu}{2\pi k_B T}\right)^{3/2} e^{-\mu v_r^2/2k_B T}\, 4\pi v_r^2\, dv_r\,. \tag{1.6}$$

This may be written, alternatively, in terms of the collision (kinetic) energy, $\epsilon_t = \frac{1}{2}\mu v_r^2$, using the relationship $d\epsilon_t = \mu v_r\, dv_r$,

$$f(\epsilon_t)\,d\epsilon_t = \frac{2}{\pi^{1/2}(k_B T)^{3/2}}\,\epsilon_t^{1/2}\, e^{-\epsilon_t/k_B T}\, d\epsilon_t\,. \tag{1.7}$$

In addition, at thermal equilibrium, the internal (rotational, vibrational, and electronic) states of the reactant molecules are populated according to the Boltzmann law,

$$p(i) = \frac{N_i}{N} = \frac{g_i\, e^{-\epsilon_i/k_B T}}{q_{\text{int}}}\,, \tag{1.8}$$

where q_{int} is the reactant molecular partition function for the internal degrees of freedom, and g_i is the degeneracy of level i at energy ϵ_i.

From state-to-state cross-sections to rate coefficients

To find an expression relating thermal rate coefficients to state-to-state reaction cross-sections we need, first, to transform the cross-sections (with dimensions area/molecule) to *state-to-state rate coefficients* (with dimensions of 1/(concentration × time)). The initial state to final state reaction rate, $\nu_{if}(v_r)$, for reactants approaching with relative velocity v_r, may be written

$$\begin{aligned}\nu_{if}(v_r) &= v_r\, \sigma_{if}(v_r)\,[\text{A}]\,[\text{BC}(i)]\\ &= k_{if}(v_r)\,[\text{A}]\,[\text{BC}(i)],\end{aligned} \tag{1.9}$$

where [BC(i)] represents the concentration (the number per unit volume) of reactant BC molecules in level i. The quantity $v_r\sigma_{if}(v_r)$ has been equated with the internal state and velocity specific rate coefficient, $k_{if}(v_r)$. Analogous to simple collision theory, $v_r\sigma_{if}(v_r)$ represents the reactive volume swept out per unit time. Unlike the simple collision theory expression, $v_r\sigma_{if}(v_r)$ refers specifically to a reaction in which the collision energy, and the initial and final states of the reactants and products, are well defined.

The *thermal* state-to-state rate coefficient, $k_{if}(T)$, which is specified at fixed temperature rather than fixed velocity or kinetic energy, is obtained by averaging the velocity specific rate coefficients in eqn 1.9 over the Maxwell–Boltzmann distribution of velocities, given in eqn 1.6,

$$\begin{aligned}k_{if}(T) &= \int_0^\infty v_r\, \sigma_{if}(v_r) f(v_r)\, dv_r\\ &= \left(\frac{\mu}{2\pi k_B T}\right)^{\frac{3}{2}} \int_0^\infty v_r\, \sigma_{if}(v_r)\, e^{-\mu v_r^2/2k_B T}\, 4\pi v_r^2\, dv_r.\end{aligned} \tag{1.10}$$

Note that the average, $\langle v_r\, \sigma_r(v_r)\rangle$, is not the same as the average relative velocity times the average cross-section, $\langle v_r\rangle\langle\sigma_r(v_r)\rangle$, as might be thought from simple collision theory.

The (state-to-state) thermal rate coefficient is therefore an average of $v_r\, \sigma_r(v_r)$ over the Maxwell–Boltzmann distribution of relative velocities, $\langle v_r\, \sigma_r(v_r)\rangle$.

State-to-state rate coefficients to thermal rate coefficients

Because the *total* removal rate from an initial state, i, is the sum of the removal rates from this state into all possible final states, initial-state-specific rate coefficients may be defined by the equation

$$k_i(T) = \sum_f k_{if}(T). \tag{1.11}$$

Finally, state-averaged thermal rate coefficients are obtained by summing over the initial states, weighted by the populations of the reactant molecules in those initial states, $p(i)$ (eqn 1.8),

$$k(T) = \sum_i p(i)\, k_i(T). \tag{1.12}$$

This is justified by noting that the reaction rate, ν, may be written

$$\begin{aligned}\nu &= k(T)\,[\mathrm{A}]\,[\mathrm{BC}] \\ &= \sum_i k_i(T)\,[\mathrm{A}]\,[\mathrm{BC}(i)] \\ &= \sum_i k_i(T)\,p(i)\,[\mathrm{A}]\,[\mathrm{BC}].\end{aligned}$$

Introduction to $N(\epsilon)$ and reaction probabilities

Section 1.3 emphasizes the role of the reaction probability on collision in determining the magnitude of the reaction cross-section and, hence, the thermal rate coefficient. Given this fundamental role, it would seem that the thermal rate coefficient should be expressible more directly in terms of reaction probabilities rather than reaction cross-sections, as was done in the preceding sections. This is indeed the case; there is an alternative expression for the thermal rate coefficient,

$$k(T) = \frac{1}{hq_\mathrm{r}} \int_0^\infty N(\epsilon)\, \mathrm{e}^{-\epsilon/k_\mathrm{B}T}\, \mathrm{d}\epsilon\,, \tag{1.13}$$

For a precise definition of the terms appearing in this expression, and a discussion of the relationship between eqn 1.13 and those of the preceding sections, see Chapter 5.

where q_r is the total reactant molecular partition function, calculated relatively easily from a knowledge of the reactant translational, rotational, and vibrational energy levels, and ϵ is the total energy (i.e. the sum of the reactant translational, rotational, and vibrational energies). The quantity $N(\epsilon)$ is known as the *cumulative reaction probability*, and is defined as the sum of reaction probabilities (lying between 0 and 1), $P_n(\epsilon)$, over all initial reactant states, n:

$$N(\epsilon) = \sum_n g_n\, P_n(\epsilon). \tag{1.14}$$

Although presented without derivation at this stage, it is possible to appreciate that eqn 1.13 has an appealing form. All the detailed information about the collision dynamics leading to reaction is contained within a single, dimensionless term, $N(\epsilon)$, the sum of reaction probabilities for each reactant state, n. If we regard different reactant states as either purely reactive ($P_n(\epsilon) = 1$) or purely unreactive ($P_n(\epsilon) = 0$), then $N(\epsilon)$ can be thought of, rather concisely, as the number of *reactive* reactant states up to an energy, ϵ.[1] Eqn 1.13 plays a central role in the development of theories of thermal reaction rates, described in the last two chapters of this book.

2 Potential energy surfaces

A prerequisite for calculating a reaction cross-section or rate coefficient is a knowledge of the forces acting on the nuclei. Such information is usually derived from a potential energy function, known as a *potential energy surface* (PES). The word 'surface' emphasizes that the potential energy function depends on the relative positions of all the nuclei involved in a reaction, and thus is generally a function of more than one parameter.

2.1 Calculation of potential energy surfaces

One difficulty associated with the theoretical study of chemical reactions is that they involve the motion of both electrons and nuclei. Fortunately, electrons are much lighter than nuclei and respond very rapidly to changes in nuclear geometry, enabling the electronic and nuclear motions to be decoupled. It is this *adiabatic* separation of electronic and nuclear motion, known as the Born–Oppenheimer approximation, which allows chemists to view reactions in terms of the motions of nuclei over single electronic potential energy surfaces.

Born–Oppenheimer separability

The use of quantum mechanics is essential for treating the electronic motion. Within the Born–Oppenheimer approximation, the total molecular wavefunction, $\Psi(r_\mathrm{e}, R_\mathrm{n})$ (which is a function of generalized electronic and nuclear coordinates, r_e and R_n) is written as a product of electronic and nuclear terms

$$\Psi(r_\mathrm{e}, R_\mathrm{n}) = \Psi_\mathrm{e}(r_\mathrm{e}; R_\mathrm{n})\, \psi_\mathrm{n}(R_\mathrm{n})$$

where $\Psi_\mathrm{e}(r_\mathrm{e}; R_\mathrm{n})$ is the wavefunction characterizing the electronic motion, and $\psi_\mathrm{n}(R_\mathrm{n})$ is that describing the nuclear motion. Ψ_e depends on the relative positions of all the electrons, r_e and, parametrically, on the nuclear positions, R_n, and satisfies the Schrödinger equation for the electronic motion:

$$\hat{H}_\mathrm{e}\, \Psi_\mathrm{e}(r_\mathrm{e}; R_\mathrm{n}) = E(R_\mathrm{n})\, \Psi_\mathrm{e}(r_\mathrm{e}; R_\mathrm{n}).$$

The resulting eigenvalues, $E(R_\mathrm{n})$, which include internuclear repulsions, correspond to the electronic energy levels, or states, of the molecule at each fixed nuclear geometry, R_n. Repeating the calculation at other nuclear geometries allows the variation of the total electronic energy with R_n to be mapped out. The function $E(R_\mathrm{n})$ may be regarded as the potential energy which the nuclei experience at a given R_n, and henceforth we will use the symbol $V(R_\mathrm{n})$ for this energy rather than $E(R_\mathrm{n})$. It is employed to solve the Schrödinger equation for the nuclear motion

$$\hat{H}_\mathrm{n}\, \psi_\mathrm{n}(R_\mathrm{n}) = \left[\hat{T}_\mathrm{n} + V(R_\mathrm{n})\right] \psi_\mathrm{n}(R_\mathrm{n}) = E_\mathrm{n}\, \psi_\mathrm{n}(R_\mathrm{n}), \tag{2.1}$$

where $\hat{T}_\mathrm{n}$ represents the kinetic energy operator for the nuclear motion. If the potential energy surface employed in this equation corresponded to that of a stable molecule, then the E_n would be the bound rovibrational energy levels of the molecule.

One important consequence of the Born–Oppenheimer approximation is that the potential energy between the atoms involved in a reaction is independent of isotopic mass. In fact, study of the variation in energy level structure of stable molecules upon isotopic substitution provides an important test of the validity of the Born–Oppenheimer approximation.

Review of molecular orbital theory

In molecular orbital theory, the electronic wavefunction for the molecule, Ψ_e, is approximated by a product of one-electron orbitals, ψ_i,

$$\Psi_e = \prod_i^n \psi_i$$

Electron spin wavefunctions should be included in the one-electron orbitals. If they are, the product is replaced by a *Slater determinant*, which ensures that the total electronic wavefunction, including electron spin, satisfies the Pauli exclusion principle.

where the product runs over all the electrons, n, in the molecule. This approximation is analogous to (and, indeed, borrowed from) the orbital approximation employed in atomic electronic structure calculations. Each electron, in its one-electron orbital, is assumed to experience an average electron–electron repulsion determined by the average positions of the other electrons in the molecule. One approach to the molecular electronic structure problem is to express the one-electron molecular orbitals as linear combinations of atomic orbitals, ϕ_j (hence the acronym LCAO approximation),

$$\psi = \sum_j c_j \phi_j , \tag{2.2}$$

where the expansion coefficient c_j is the amplitude of the jth atomic orbital in the one-electron molecular orbital, ψ.

The wavefunctions given above are approximate ones, and a procedure is required to optimize them, and determine the set of coefficients c_j which lead to the best approximation to the true wavefunction. One commonly used method employs the *variational principle*, which states that the approximate ground state energy of a molecule, E_{trial}, obtained using a trial wavefunction, Ψ_{trial}, will always be an upper bound to the true ground state energy of the molecule, E:

$$E_{\text{trial}} = \frac{\int \Psi^*_{\text{trial}} \hat{H} \Psi_{\text{trial}}}{\int \Psi^*_{\text{trial}} \Psi_{\text{trial}}} \geq E .$$

In simple molecular orbital theory, this procedure is employed to find the best set of coefficients (i.e. the set which minimizes the energy, E_{trial}) by setting the derivatives of E_{trial} with respect to each of the coefficients, c_j, to zero. This leads to a set of simultaneous equations, known as the *secular equations*, the solution of which yields the coefficients and the orbital energies of interest.

For many-electron problems, the trial ground state wavefunction of the molecule must be optimized iteratively, since the average electron–electron repulsion experienced by one electron, in its approximate one-electron orbital, depends on the one-electron wavefunctions of all the other electrons in the molecule. The method frequently employed for this optimization is known as the self-consistent field procedure. However, even this methodology fails to take proper account of electron–electron repulsion, since it neglects the instantaneous correlations between the positions of each of the electrons in the molecule. These effects may be accounted for using techniques such as configuration interaction, and more details of this and other procedures can be found in texts given in the Background reading.

To calculate a potential energy *surface*, the Schrödinger equation for the electronic motion must be solved many times, over a wide range of nuclear configurations, R_n. This yields the potential energy at a series of points, corresponding to the different nuclear geometries, and these are usually fitted with some analytic function so that potential energy may be evaluated at any arbitrarily chosen geometry. The latter procedure is necessary for reaction dynamics calculations, which require the potential energy to be evaluated continuously from reactants to products. To date, global *ab initio* potential energy surfaces of sufficient accuracy to provide reliable rate coefficient data have only been obtained for comparatively small systems (both in terms of the number of atoms and the number of electrons), such as $H + H_2$, $F + H_2$, and $H + H_2O$.

For the four-atom system, most of the computational effort in calculating the potential energy surface has focused on the region close to the minimum energy path from reactants to products (see Section 2.2), and the surface is only believed accurate for modelling the $H + H_2O \longrightarrow OH + H_2$ reaction at low energy.[2]

Because of the difficulty in calculating accurate *ab initio* potential energy surfaces for larger systems than those described above, numerous approximate procedures for estimating features of the potential energy surface have been developed. One commonly employed strategy is to rescale the reactant and product energies (and sometimes the vibrational frequencies) obtained from low level *ab initio* calculations (i.e. calculations which do not account properly for electron correlation, or employ small basis sets in the expansion of the molecular orbitals given in eqn 2.2) to those determined experimentally. Another is to calculate the potential energy accurately only in regions that are thought to be important in the reaction, for example in the region of the potential energy barrier (see below). Even more approximate are the LEPS (London, Eyring, Polanyi, and Sato) and DIM (diatomics in molecules) *semi-empirical* potential energy surfaces, details of which can be found in the texts given in the Background reading. Both of these surfaces are constructed solely from experimentally derived knowledge of the diatomic potential energy curves for the species AB, BC, and AC. One important feature of the LEPS surface is that it is expressed by a comparatively simple analytic function, which enables the potential energy to be evaluated efficiently at any nuclear geometry.

2.2 Types of potential energy surface

Diatomic potential energy curves

For a diatomic molecule, the potential energy *curve* is a function of a single parameter, the bond length or atomic separation, R. For a stable molecule, it has a familiar Morse-like form, shown in Fig. 2.1. Also included in this figure is a repulsive excited state of the molecule, which might arise from promotion of a valence electron from a bonding to an antibonding orbital. The potential energy curve for the stable ground electronic state is repulsive at short range, but becomes attractive at longer range, tending to the dissociation limit of the two separated atoms (the potential energy in this limit would generally be set to zero, if ground electronic state fragments were produced in the dissociation process). This potential curve supports bound vibrational levels (i.e. the solutions to the Schrödinger equation for the nuclear motion, eqn 2.1) up to the dissociation limit, where they converge to the dissociation continuum. Approximate, harmonic vibrational frequencies,

$V(R)$

D_e

R

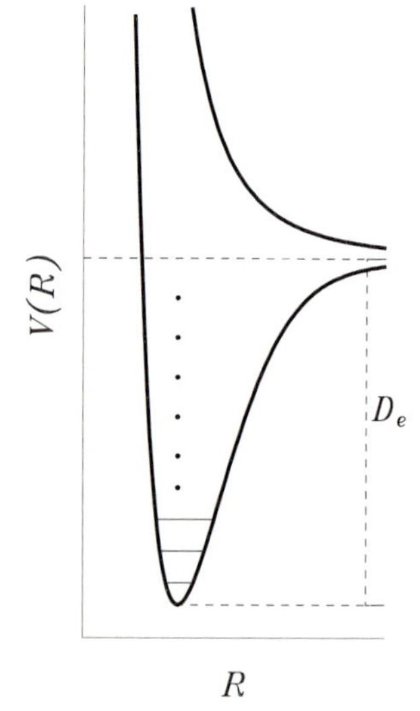

Fig. 2.1 Schematic potential energy curves for a diatomic molecule. D_e is the dissociation energy measured from the bottom of the well.

$$\nu = \frac{1}{2\pi}\sqrt{\frac{k}{\mu}},$$

can be obtained from the potential energy curve by determining the force constant, k, defined as

$$k = \left(\frac{d^2V}{dx^2}\right)_{x=0},$$

where x is the displacement $(R - R_e)$, and R_e is the equilibrium bond length of the molecule at the bottom of the potential energy well; this bond length also serves to define the equilibrium rotational constant of the molecule. Most pertinent of all, knowledge of the potential energy curve also allows determination of the force on the atoms at any internuclear separation, via the equation,

$$F(R) = -\frac{dV(R)}{dR}.$$

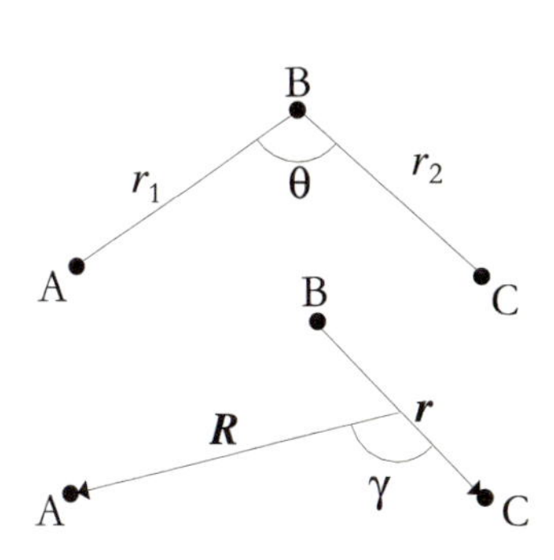

Fig. 2.2 Two choices of coordinates used to describe a triatomic PES.

If the two atoms involved were closed-shell, noble gas atoms, then the form of the potential energy curve would be rather different from the Morse-like behaviour observed in stable diatomic molecules. Under these circumstances, the potential at short range is dominated by the repulsive interaction between the nuclei and between the two filled charge clouds (cf. the repulsive curve shown in Fig. 2.1). At long range, however, the potential exhibits a shallow well, associated with long range attractive, dispersion interactions, known as *van der Waals* forces. The potential energy curves for weakly bound van der Waals complexes are difficult to calculate accurately using *ab initio* techniques, because the origin of the attractive forces lies in the instantaneous correlation between the positions of the electrons in the two atoms. It is this electron correlation energy which is most difficult to allow for in potential energy surface calculations. The potential curves are often modelled using a function known as the Lennard–Jones potential,

$$V(R) = 4\epsilon\left[\left(\frac{\sigma}{R}\right)^{12} - \left(\frac{\sigma}{R}\right)^{6}\right], \qquad (2.3)$$

where ϵ is the well depth and σ the separation when the potential energy is zero.

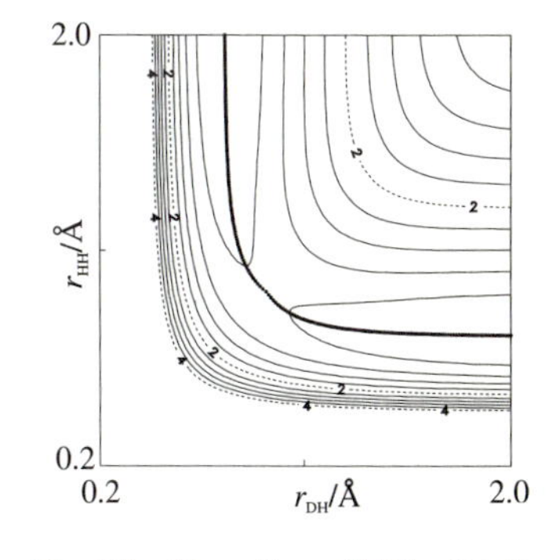

Fig. 2.3 The collinear PES for $D + H_2$.

Triatomic potential energy surfaces

The potential energy surface for a three-atom system is a function of three coordinates, either the three interatomic bond lengths or, more commonly, the bond lengths of BC and AB, r_{BC} and r_{AB}, and the bond angle between the BC and AB bonds, θ (see Fig. 2.2). To start with, we will use the latter. To plot the potential energy surface, one of the degrees of freedom needs to be frozen and we will fix the bond angle to 180° (such that A approaches BC in a *collinear* fashion). The potential energy can then be plotted as a function of the two bond lengths.

A discussion of three-atom systems in terms of simple MO theory can be found in Levine and Bernstein's *Molecular reaction dynamics and chemical reactivity.*

In general, for an N-atom reaction, the potential energy surface is a function of $3N - 6$ degrees of freedom.

An example of the resulting *collinear* potential surface is illustrated in Fig. 2.3 for the $D + H_2$ reaction (we use the D isotope merely to distinguish the different atoms).[3] The surface is mapped in the form of a contour plot, where the contours connect points of equal potential energy. Within the Born–Oppenheimer approximation, the potential energy of the reactants (arbitrarily taken as zero) is identical to that of the products, and thus the diagram is symmetric about the diagonal line $r_{HD} = r_{HH}$. Consider taking a slice through

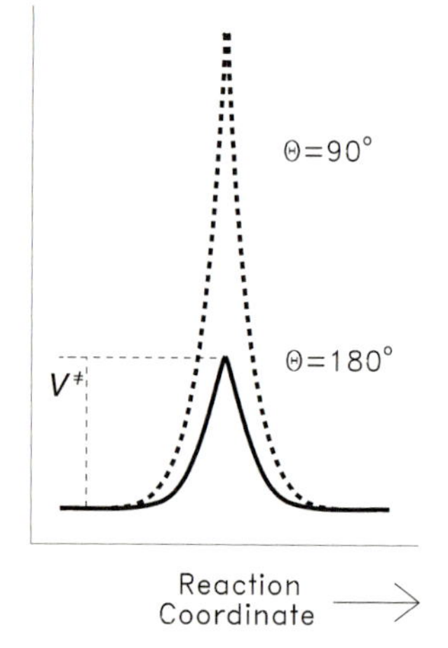

Fig. 2.4 The variation in potential energy along the MEP for $D + H_2$. The upper curve shows the analogous variation for $\theta = 90°$. $V^{\neq}$ is the potential energy barrier height.

For the isotopically pure species, H′–H–H, the symmetric stretching coordinate lies along the line $r_{H'H} = r_{HH}$ in Fig. 2.3. The dissociation limit reached by stretching motion orthogonal to the MEP therefore corresponds to the completely separated atoms, $H' + H + H$.

the potential parallel with the r_{HH} axis when r_{DH} is large and constant (i.e. parallel to the *y* axis at the extreme right of the diagram). This region of the potential corresponds to the reactants, and the potential energy curve along the slice is simply the diatomic Morse-like potential of the H_2. A similar slice through the potential at large, fixed r_{HH} separations (corresponding to a line at the top of the diagram, parallel to the *x* axis), which is in the product region of the potential energy surface, yields the diatomic potential curve of DH. Reaction takes place when the two reactants come together, and the region of the surface where all three atoms are in close proximity is the bottom left of the diagram (conversely, the top right of the diagram represents the potential energy when the three atoms become separated).

The bold line in Fig. 2.3 represents the lowest energy path from reactants to products, and is known as the *minimum energy path* (MEP) or the *reaction coordinate*. The variation of the potential energy along this coordinate is illustrated in Fig. 2.4. As the D atom approaches H_2, the potential energy rises along the *reactant valley* until the barrier is reached, and subsequently falls as the D–H–H species passes into the *product valley* of the surface. Motion along the MEP is best viewed as a translational motion over a barrier, taking DH_2 from the reactants into the products. By contrast, motion at right angles (orthogonal) to the MEP represents the symmetric stretching *vibrational* coordinate of the D–H–H species. The variation of potential energy along this coordinate corresponds to a well rather than a barrier, and is similar in form to the diatomic Morse-type curve illustrated in the previous section. Therefore, the barrier region encompasses a *saddle point* on the potential energy surface. Motion along the reaction coordinate corresponds to unbound, translational motion over a barrier, while motions along other coordinates, orthogonal to the MEP, correspond to bound vibrational motions.

If we are to describe reactions properly, we must allow for D atom collisions which approach H_2 in a sideways fashion, as well as collisions involving rotationally excited H_2 reactants and DH products. To account for such events, knowledge of the variation of potential energy with bond angle is essential. In addition to the collinear potential energy surface shown in Fig. 2.3, a whole family of similar surfaces must be generated, corresponding to different bending angles in the DH_2 species. For the DH_2 system, detailed calculations show that the potential energy rises steeply as the bond angle is reduced from 180°, and thus the minimum energy path for this reaction corresponds to the collinear path already shown in Fig. 2.4. This diagram also shows the variation in the potential energy in the barrier region as the bond angle is reduced from 180°. The bond lengths employed to illustrate this variation are the same as those used to define the MEP.

A rather different potential energy surface is obtained for the $F + H_2$ system, shown for collinear approach in Fig. 2.5,[4]

$$F + H_2 \longrightarrow HF + H.$$

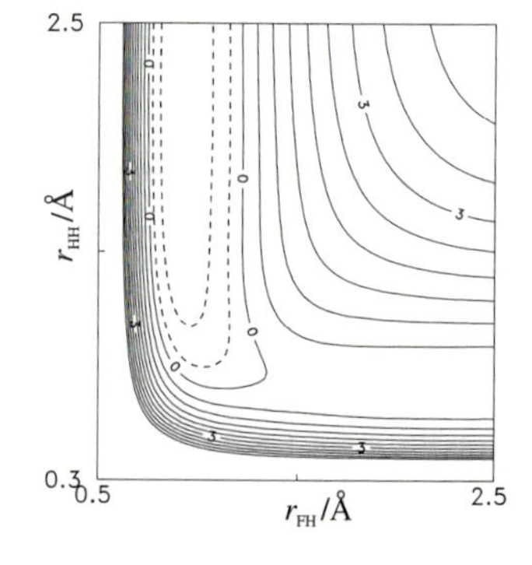

Fig. 2.5 The collinear PES for $F + H_2$.

Since the HF bond is much stronger than that of H_2, the reaction is highly exothermic. For clarity, contours with energy greater than zero (defined here as the energy of $F + H_2$ at infinite separation) are shown as continuous lines, while negative potential energy contours are shown as dashed lines. The potential energy surface is no longer symmetric, and the modest barrier to reaction occurs in the entrance, reactant valley. Such a surface is said to possess an *early barrier*. Another interesting feature of the surface is that the minimum energy path for this reaction does not correspond to collinear approach. To illustrate this, it is convenient to employ an alternative set of coordinates, R, r and γ, the separation of F to the H_2 centre-of-mass, the H–H internuclear separation, and the angle between the vector $\mathbf{R}$ and the H_2 bond axis, $\mathbf{r}$ (see Fig. 2.2). This representation of the surface is shown in Fig. 2.6. The remaining coordinate, the H_2 bond length, r, has been fixed at its optimized value at the saddle point ($r = 0.77$Å).

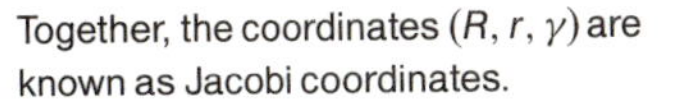
Together, the coordinates (R, r, γ) are known as Jacobi coordinates.

Our last example of a potential energy surface is that for the reaction of electronically excited oxygen atoms, $O(^1D)$, with H_2,[5]

$$O(^1D) + H_2 \longrightarrow OH + H.$$

This reaction proceeds over a surface with a deep potential energy minimum, corresponding to the ground electronic state of the water molecule, and the minimum energy path corresponds to insertion of the O atom into the H_2 bond, with a perpendicular angle of approach. It is convenient to employ the same set of coordinates as used in the previous example, and in Fig. 2.7 the potential energy surface is plotted as a function of R and r, the O–H_2 separation and the H_2 bond length. The angle, γ, between the vectors $\mathbf{R}$ and $\mathbf{r}$ has been fixed at 90°; the point group of the HOH species corresponding to this diagram is, therefore, C_{2v}. The potential energy minimum on the surface, which corresponds to the ground electronic state of the water molecule, is also shown in the figure. Note that there is no barrier to $O(^1D)$ insertion on the surface.

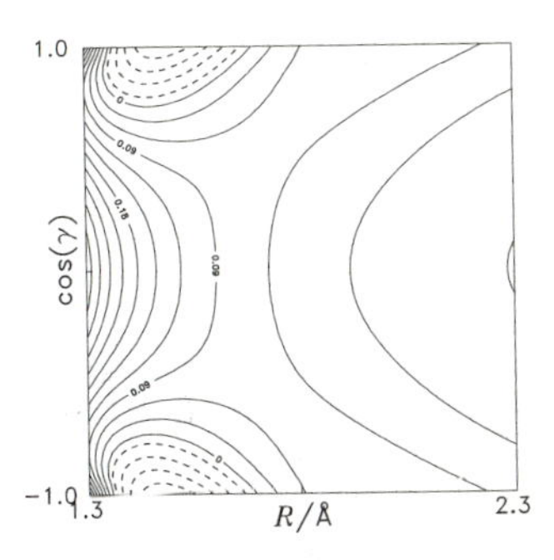

Fig. 2.6 The dependence of the PES for $F + H_2$ on Jacobi coordinates R and γ.

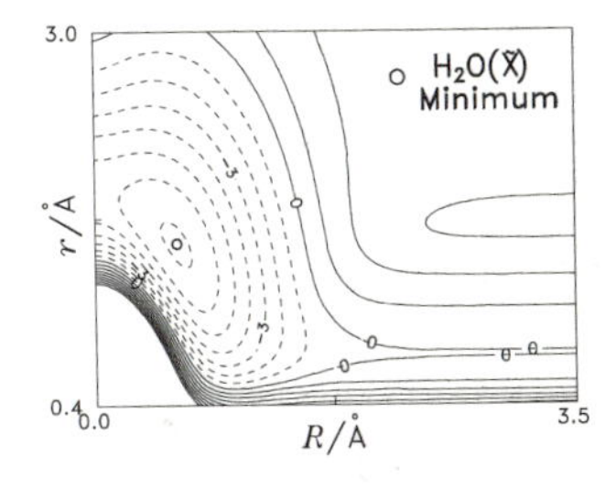

Fig. 2.7 The R, r dependence of the PES for $O(^1D) + H_2$. The Jacobi angle γ is fixed at 90°.

Non-adiabaticity and correlation rules

Thus far we have assumed that the Born–Oppenheimer approximation is valid. However, there are occasions when it is not possible to separate nuclear and electronic motion, particularly at energies close to molecular dissociation limits, where, within the Born–Oppenheimer picture, several electronic states often converge to a common dissociation asymptote. In such circumstances, allowance must be made for the coupling between the nuclear and electronic motions, and this is often achieved by treating the coupling as a perturbation to the decoupled, Born–Oppenheimer motion.

A concise account of the treatment of non-adiabatic effects is given by Billing and Mikkelsen in *Molecular dynamics and chemical kinetics*.

Examples of non-adiabatic effects arise in the chemical reactions between alkali metal atoms with halogen molecules, such as

$$K + I_2 \longrightarrow KI + I,$$

which at high collision energy undergoes a competing electron transfer reaction to yield $K^+ + I_2^-$. Reaction to produce neutral products proceeds via a mechanism known as the *harpoon* mechanism, which involves an electron jump (the harpoon) from the K atom to the I_2 molecule at large K–I_2 separations. The electron transfer occurs at the intersection between the covalent potential energy surface, which is largely repulsive in character, and an attractive ionic surface, correlating with the $K^+ + I_2^-$ asymptote. Because electron transfer takes place at large internuclear separations, these reactions usually possess very large reaction cross-sections and rate coefficients (see Section 4.2). The generation of ionic products at high collision energies reflects the crossing to the ionic surface, and the production of ions with sufficient energy to escape the Coulomb attraction.

Another example of such a reaction was provided in Table 1.1, namely that of $K + Br_2$. Note that the reaction possesses a very large Arrhenius A-factor, and a steric factor exceeding unity.

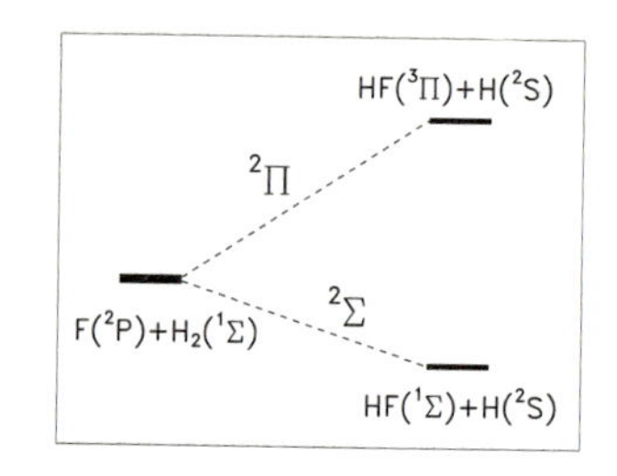

Fig. 2.8 Correlation diagram for $F + H_2$.[6]

One means of ascertaining whether non-adiabatic effects are likely to be important is to identify the electronic symmetries of the low-lying electronic states of reactant and product species, and attempt to correlate these states according to their symmetries. An example of such a correlation diagram for the $F + H_2$ reaction is shown in Fig. 2.8.[6] For simplicity, the diagram is drawn assuming collinear approach of F to H_2. The ground state of the F atom is ^2P, and the *p* orbital containing the unpaired electron may approach either along the H–H axis or perpendicular to it. These two configurations lead to $^2\Sigma$ and $^2\Pi$ states of the F–H–H supermolecule. Switching attention to the products, it is known from simple molecular orbital (MO) considerations that HF has a $^1\Sigma$ ground state and a first excited $^3\Pi$ state. When H(^{2}S) atoms approach the HF molecule in either of these two states (again in a collinear configuration) the resulting doublet F–H–H molecular states produced are of $^2\Sigma$ and $^2\Pi$ symmetry. Thus, the lowest electronic states in both the reactant and the product regions are of $^2\Sigma$ symmetry and correlate adiabatically with one another on a single potential surface. This suggests that non-adiabatic effects in the $F + H_2$ reaction are unlikely to play a dominant role.

Further examples of correlation diagrams can be found in Herzberg's *Electronic spectra and electronic structure of polyatomic molecules.*

2.3 Experimental probes of potential energy surfaces

The precise definition of the term *transition state* is given in Chapter 5. For now we simply refer to the barrier region of the potential energy surface, where the two reactant species are in close proximity.

Spectroscopic techniques for probing transitions between electronic, vibrational, and rotational energy levels of stable diatomic and polyatomic molecules are well established. It would be desirable to have similar spectroscopic tools to probe regions of the potential energy surface well away from the reactant and product asymptotes, close to the barrier or *transition state region.*[7]

The barrier region of the potential energy surface does not correspond to a stable molecule, which makes the direct spectroscopic interrogation of the transition state region very difficult. One method of stabilizing the (reactive) reactant molecules makes use of the long range attractive van der Waals forces, which can lead to a small potential well at large internuclear separations, even for reactions proceeding over surfaces with a barrier. The reactants are cooled by adiabatic expansion of a high pressure of the reactant mixture through a pin-hole into a low pressure chamber, a technique known as *jet cooling*. Collisions immediately downstream of the expansion orifice lead to the formation and stabilization of van der Waals complexes, which are then further cooled by subsequent energy transferring collisions. Once generated, the van der Waals molecules can be probed by a variety of spectroscopic techniques. Examples are provided by studies of Ca–HX van der Waals complexes. The electronic spectra of the complexes provide information about transition state regions of the potential energy surfaces for the reactions of electronically excited Ca* atoms with the hydrogen halides (HX).[8]

An introduction to photoelectron spectroscopy can be found in Atkins' *Physical chemistry.*

A recent alternative spectroscopic method for probing the transition state region employs a variant of photoelectron spectroscopy known as anion photoelectron detachment spectroscopy. In these experiments, stable though weakly bound anions of the species of interest, such as FH_2^- and IHI^-, are made by electron bombardment of a suitable mixture of jet cooled precursor molecules. Once generated, the anions are studied by conventional photoelectron spectroscopy, which involves measuring the kinetic energy distribution of the electrons ejected by photon absorption, for example,[9]

$$FH_2^- \xrightarrow{h\nu} FH_2 + e^-.$$

The species left behind after electron ejection is the neutral 'supermolecule' (FH_2), rather than a positive ion, as is the case in conventional photoelectron spectroscopy. The distribution (or spectrum) of electron kinetic energies reflects the distribution of internal rotational and vibrational energies, E_{int}, accessed in the neutral species. This follows from the simple energy conservation equation for the electron kinetic energy,

$$E_t = h\nu - E_{int} - D_0 - E_{ea},$$

where D_0 is the dissociation energy of the anion and E_{ea} is the electron affinity of F, as shown in Fig. 2.9.

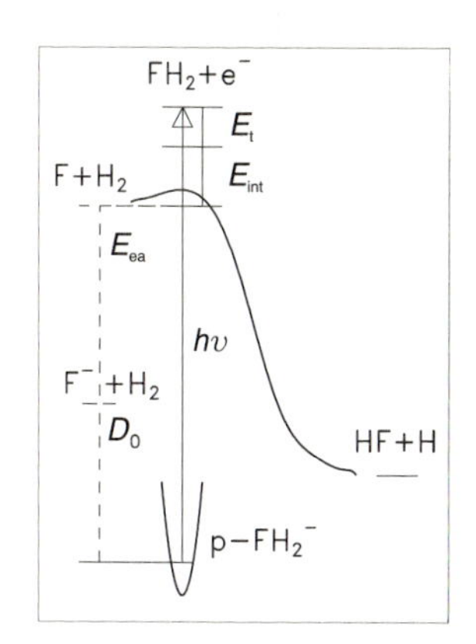

Fig. 2.9 Schematic diagram of the anion photoelectron detachment experiment on FH_2^-.

The resulting photoelectron spectrum of FH_2^- is shown in Fig. 2.10. It consists of a series of broad transitions, corresponding to excitation to vibrational 'states' in the barrier region of the F + H_2 reaction. The broadening of the lines reflects the *lifetime broadening* associated with rapid dissociation of FH_2 to F + H_2 and to HF + H. The transitions themselves can be rationalized using the same Franck–Condon arguments employed to discuss electronic and photoelectron spectra of diatomic molecules. *Ab initio* potential energy surface calculations suggest that the anion, FH_2^-, has a stable linear geometry (with the F^- anion on average sitting about 2 Å from the centre-of-mass of H_2), whereas the neutral FH_2 supermolecule has a preferred bent configuration in the barrier region (see Fig. 2.6). The spectral features of this linear-to-bent transition thus correspond to excitation to bending levels of FH_2. These bending levels are best thought of as hindered rotations of H_2 about the F–H_2 bond, which become excited when the electron on the linear anion is suddenly removed in the photodetachment process. The spectrum thus provides very direct experimental evidence that the minimum energy path for the F + H_2 reaction passes through a bent FH_2 structure at the barrier.

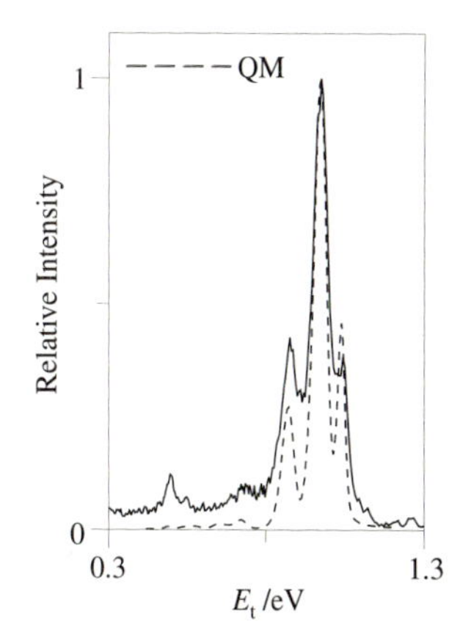

Fig. 2.10 The anion photoelectron detachment spectrum of FH_2^-. The dashed curve is the quantum calculated spectrum.

2.4 Motion over the surface

Motion of the nuclei over the potential energy surface cannot be regarded simply as translation along a one-dimensional minimum energy path. The motion is a multidimensional one, and the method chosen to determine the nuclear dynamics depends on the experimental property being modelled, as well as the complexity of the reaction to be studied. Ideally quantum mechanical methods should be employed, but this is still only feasible for relatively light three- or four-atom systems. Frequently, classical mechanics is employed as an approximate alternative. The validity of employing classical mechanics for reaction dynamics studies is still debated, and aspects of this issue will be touched on henceforward.

Classical mechanics

The classical motion of the atoms involved in a collision is represented by the time dependence of their positions, usually referred to as a *trajectory*. The motion is determined by solving Newton's laws, subject to a set of initial conditions, which correspond to the initial velocities or *momenta* of each reactant atom and their initial relative positions. Newton's laws of motion are usually expressed in terms of Hamilton's equations,

A commonly employed variant of the *classical trajectory* method is the *quasi-classical trajectory* (QCT) method, in which the initial rotational and vibrational energies of the reactants are chosen to correspond to the known quantum mechanical energy levels of the molecules (see Section 3.2).

$$\frac{\partial H}{\partial P_i} = \frac{\mathrm{d}Q_i}{\mathrm{d}t} \equiv \dot{Q}_i$$
$$-\frac{\partial H}{\partial Q_i} = \frac{\mathrm{d}P_i}{\mathrm{d}t} \equiv \dot{P}_i, \tag{2.4}$$

where Q_i represents the ith (X, Y, and Z) coordinate, with associated momentum P_i, and H, known as the Hamiltonian, is the total energy of the system (i.e. the sum of the kinetic, T, and potential, V, energies, expressed in the above position and momentum 'coordinates'). For a reaction involving three atoms there will be a total of $3N = 9$ coordinates and nine momenta, corresponding to the positions and momenta of each atom along the X, Y, and Z axes. Thus, for three atoms, Hamilton's equations of motion (2.4) are a set of 18 coupled differential equations. As shown below, six of these equations describe the position and momentum of the centre-of-mass of the system, which is conserved throughout the reaction, and can therefore be factored out of the problem. This leaves a total of 12 differential equations which must be solved numerically for each set of initial conditions to characterize the time evolution (trajectory) of the three atoms. The trajectories must satisfy the laws of conservation of energy and (linear and angular) momentum, and these constraints are normally employed as a check that the numerical integration has been performed with sufficient accuracy.

To calculate a reaction cross-section or rate coefficient, many trajectories (typically $> 10^3$) must be calculated, each trajectory corresponding to a different initial condition of the system. The sampling of the latter must reflect the layers of averaging described in Chapter 1, although the precise details of the sampling depend on the property to be determined (see Chapter 3).

The application of Hamilton's equations can be illustrated in the simple case of a collision between two atoms, labelled A and B. The Hamiltonian, the sum of the kinetic and potential energies of the two particles, is most conveniently written

$$H = \frac{\mathbf{P}_{\mathrm{cm}}^2}{2M} + \frac{\mathbf{P}^2}{2\mu} + V(R), \tag{2.5}$$

where the terms appearing in the expression are defined explicitly as

$$\mathbf{P}_{\mathrm{cm}} = M\dot{\mathbf{R}}_{\mathrm{cm}}$$
$$\mathbf{P} = \mu\dot{\mathbf{R}}$$
$$\mathbf{R}_{\mathrm{cm}} = \frac{1}{M}(m_{\mathrm{A}}\mathbf{R}_{\mathrm{A}} + m_{\mathrm{B}}\mathbf{R}_{\mathrm{B}})$$
$$\mathbf{R} = \mathbf{R}_{\mathrm{A}} - \mathbf{R}_{\mathrm{B}}$$

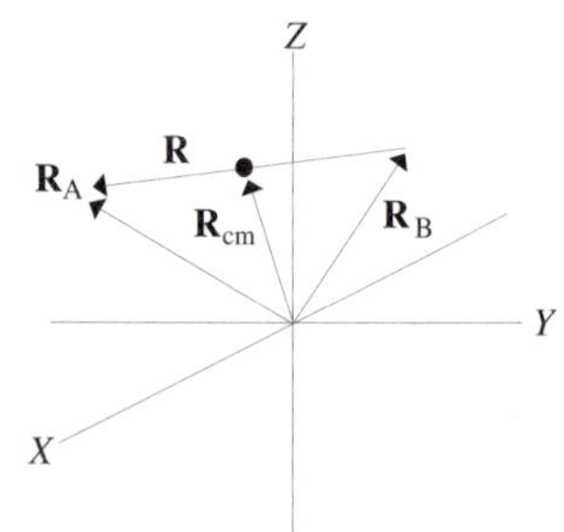

Fig. 2.11 Coordinates for motion of two atoms.

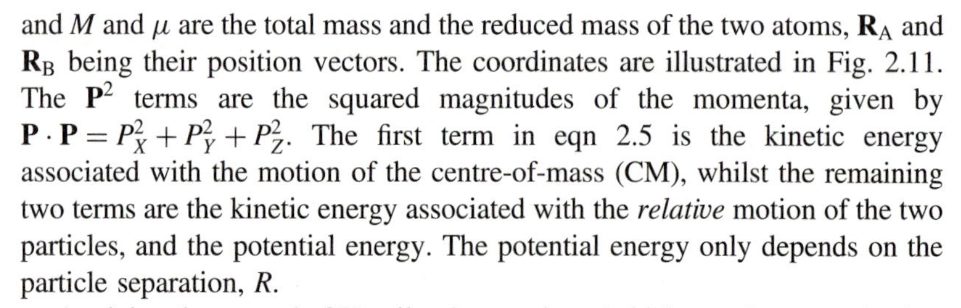

and M and μ are the total mass and the reduced mass of the two atoms, $\mathbf{R}_{\mathrm{A}}$ and $\mathbf{R}_{\mathrm{B}}$ being their position vectors. The coordinates are illustrated in Fig. 2.11. The $\mathbf{P}^2$ terms are the squared magnitudes of the momenta, given by $\mathbf{P} \cdot \mathbf{P} = P_X^2 + P_Y^2 + P_Z^2$. The first term in eqn 2.5 is the kinetic energy associated with the motion of the centre-of-mass (CM), whilst the remaining two terms are the kinetic energy associated with the *relative* motion of the two particles, and the potential energy. The potential energy only depends on the particle separation, R.

Applying the second of Hamilton's equations (which may be recognized as an expression for Newton's second law) to each of the coordinates of the centre-of-mass motion, we obtain the following equation:

$$-\frac{\partial H}{\partial R_{\mathrm{cm},\,i}} \equiv -\frac{\partial V}{\partial R_{\mathrm{cm},\,i}} = 0 = \frac{\mathrm{d}P_{\mathrm{cm},\,i}}{\mathrm{d}t}.$$

The last equality states that the momentum (and hence the velocity) of the centre-of-mass does not change with time; i.e. that it is a constant during the collision. This is because there is no *external* force ($= -\partial V/\partial R_{\mathrm{cm}}$) acting on the centre-of-mass: the potential energy only depends on the interatomic separation (i.e. the *relative* positions of atoms A and B). Thus, we can write the Hamiltonian describing the relative motion of the two atoms as

The frame of reference (i.e. the coordinate axes) employed to describe the relative motion of the particles is usually referred to as the centre-of-mass (CM) frame. In this frame, the position of the CM of the particles is stationary during the collision encounter, and the CM linear momenta of the two particles are equal and opposite at all times.

$$H = \frac{\mathbf{P}^2}{2\mu} + V(R), \tag{2.6}$$

where $\mathbf{P}^2$ is the squared magnitude of the momentum associated with the relative motion of the two particles. Although this example is a simple one, the final result applies more generally. All reactions can be described using a simplified Hamiltonian written in terms of relative positions and momenta.

Quantum calculations

Major advances in quantum dynamical calculations have been made in the past two decades, facilitated by new efficient numerical algorithms and the dramatic improvements in computer speed and memory capacity. The problem of solving the Schrödinger equation for the nuclear motion scales not just with the number of atoms (as is the case with Newton's laws), but also with the number of reactant and product quantum states accessible at a given energy or temperature. Thus, although there have been several examples of accurate quantum dynamics calculations of three-atom reactions in which all degrees of freedom have been treated rigorously, most work has been performed on direct reactions involving light atoms. Reactions proceeding over surfaces with deep potential energy wells, such as the $O(^1D) + H_2$ reaction (see Section 2.1), have only recently received attention, concurrently with the study of direct four-atom reactions such as $H + H_2O$. However, for these more difficult reactions, angular momentum effects have yet to be treated exactly.

A detailed account of quantum mechanical reaction dynamics (or quantum reactive scattering theory) is beyond the scope of this book and the reader is referred to texts given in the Background reading.

The ability to solve rigorously the nuclear dynamics problem allows comparison to be made with rate coefficients and cross-sections calculated using more approximate methods employing the same potential energy surface. Although the surface may not be accurate, such a comparison of results enables chemists to test the validity of approximate treatments. To illustrate this point, we compare experimental[10] and calculated rate constants for the reaction $D + n\text{-}H_2$ (where the n stands for normal hydrogen) obtained using exact quantum mechanical (QM)[11] methods and the quasi-classical trajectory (QCT) method[12] (see Fig. 2.12), employing the same *ab initio* potential energy surface. The calculations of both the surface and the dynamics are completely parameter-free. The agreement between the quantum results and experiment is impressive, though not perfect at high energies. Somewhat more surprising is the good correspondence between the QCT calculated and the quantum mechanical rate coefficients, with discrepancies ranging from a factor of three at low temperatures to less than a factor of two at high temperatures.

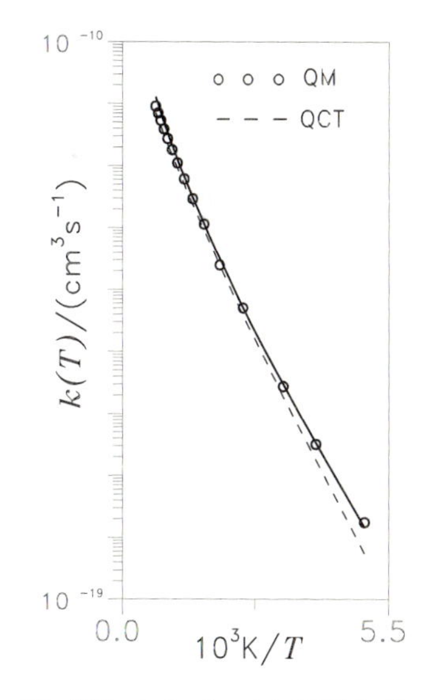

Fig. 2.12 Arrhenius plot of $k(T)$ for the $D + n\text{-}H_2$ reaction, comparing experimental, QCT, and QM rate data.

Such good agreement between *ab initio* theory and experiment is currently achievable only for the simplest of chemical reactions. Also, the good agreement between quantum and classical mechanics may not be realized for other reactions.

3 The differential cross-section

3.1 Elastic scattering

Non-reactive collisions may be categorized into those which conserve kinetic energy (elastic collisions) and those which do not (inelastic collisions). Inelastic collisions lead to the transfer of energy between translational and internal (electronic, vibrational, and rotational) degrees of freedom, while elastic collisions merely transfer kinetic energy from one species to another. We shall focus here on the latter process in order to introduce the important concepts of centrifugal kinetic energy and orbital angular momentum, which play a key role in chemical reactions.

The reader is referred to the Background reading for more information about inelastic scattering.

Experimental considerations

The beam of atoms can be generated using the methods described in Section 3.2.

The differential cross-section characterizes the *scattering angle* dependence of the collision or reaction cross-section. Imagine a beam of atoms, A, impinging on target atoms, B, as illustrated in Fig. 3.1. Assume that the target atoms are stationary, and that the A atoms are travelling at a well-defined velocity, equivalent to the relative velocity (because $v_r = v_A - v_B$, and $v_B = 0$). We will use the occurrence of a deflection in the path of atom A as a means of defining whether or not a collision has taken place. Deflections are detected by monitoring the atoms (by mass spectrometry, for example) as a function of angle with respect to the initial atomic beam direction, as shown. To define this scattering angle precisely, a collimating slit is placed in front of the detector.

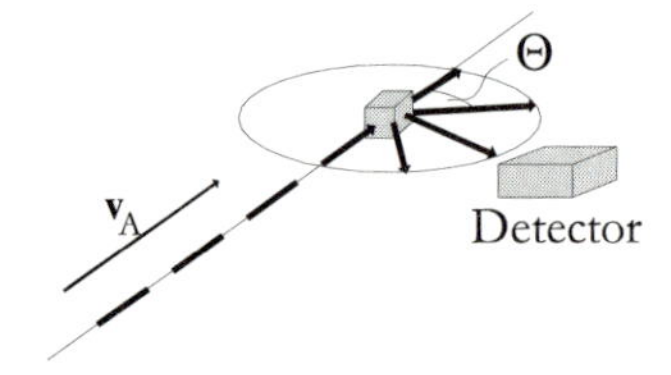

Fig. 3.1 Illustration of the measurement of the differential cross-section.

The differential cross-section may be expressed quantitatively in terms of the *flux* of scattered atoms (the number of scattered atoms arriving per unit time per unit area or per steradian) by the following equation:

$$J_A = \frac{d\sigma}{d\Omega} v_r [A] \Delta V [B], \tag{3.1}$$

The capital Ω is used to denote the polar angles (Θ, Φ) defined in the laboratory, which are not the same as those employed to characterize the *relative motion* of the particles. This suggests that it is necessary to transform experimental data from the *laboratory* frame to the *centre-of-mass* (CM) frame (see Section 3.2).

where v_r[A] corresponds to the *incident flux* of A atoms, impinging on a small target volume, ΔV, of B atoms, at concentration [B]. Note that the right side has the correct dimensions of number per unit time per steradian. Thus, with a knowledge of the beam velocity, the concentrations of atoms A and B, and the intersection, or interaction, volume, we can transform the detected signal intensity into the differential cross-section of interest.

In practice, it is not an easy task to determine all the above experimental parameters precisely and often only the *angular distribution* of the scattered products is determined. In the CM frame (in which the angular coordinates are denoted ω), we may write

$$P(\theta, \phi) = \frac{1}{\sigma} \frac{d\sigma}{d\omega},$$

where $P(\theta, \phi)$ represents the *probability density* of finding products scattered at polar angles θ, ϕ. Measurement of the angular distribution does not require a knowledge of the reactant atom concentrations or the intersection volume, but it is still desirable to perform the experiment at a well-defined relative velocity and, hence, collision energy.

Classical elastic scattering of two atoms

The classical motion of two structureless atoms is fully characterized by Hamilton's equations of motion, employing the Hamiltonian for the *relative* motion given in eqn 2.6. Because the Hamiltonian represents the total energy associated with the relative motion of the system, which is conserved during the collision, we may equate the Hamiltonian with the *initial relative kinetic energy* of the particles, ϵ_t. This follows from the fact that, initially, when the particles are at infinite separation, the potential energy may be taken to be zero:

More quantitative aspects of the classical mechanical elastic scattering are described in Appendix A.1.

$$H = \epsilon_t = \tfrac{1}{2}\mu v_r^2,$$

where $\mu = m_A m_B / m_{AB}$. Eqn 2.6 can be rewritten, more revealingly, in polar coordinates, as

$$\epsilon_t = \tfrac{1}{2}\mu \dot{R}^2 + \frac{L^2}{2\mu R^2} + V(R). \tag{3.2}$$

The first term in this equation is the kinetic energy associated with relative motion of the two atoms towards each other. It is known as the *radial kinetic energy*. The second term, referred to as the *centrifugal kinetic energy*, is the energy associated with the orbital motion of the particles, with *orbital angular momentum L*. Both the direction and magnitude of the orbital angular momentum are conserved (i.e. are constant) during an elastic collision between atoms, and the magnitude of the orbital angular momentum may be defined as

The conservation of orbital angular momentum allows us to view the atom–atom collision as occurring in a plane (see Appendix A.1).

$$L = \mu R^2 \dot{\vartheta} = \mu v_r b, \tag{3.3}$$

where b is the impact parameter for the collision introduced in Section 1.3.

The two equations of motion, eqns 3.2 and 3.3, characterize the radial and angular relative motion of the two atoms. Rearranging the latter equation yields an expression for the time dependence of the angle, ϑ

$$\dot{\vartheta} = v_r \frac{b}{R^2},$$

while the radial equation may be written

$$\tfrac{1}{2}\mu \dot{R}^2 = \epsilon_t - V_{\text{eff}}(R), \tag{3.4}$$

where $V_{\text{eff}}(R)$ is defined as

$$V_{\text{eff}}(R) = V(R) + \epsilon_t \frac{b^2}{R^2}.$$

Thus, the radial motion of the two atoms may be regarded as the motion of a particle of effective mass, μ, over an *effective potential*, $V_{\text{eff}}(R)$, which comprises the sum of the potential energy and the centrifugal kinetic energy. The latter term increases as the separation, R, decreases. The effective potential is illustrated in Fig. 3.2 for different values of the impact parameter, b, and, hence, the orbital angular momentum (Fig. 3.2). Even in the absence of a barrier in the potential energy curve, $V(R)$, the effective potential often

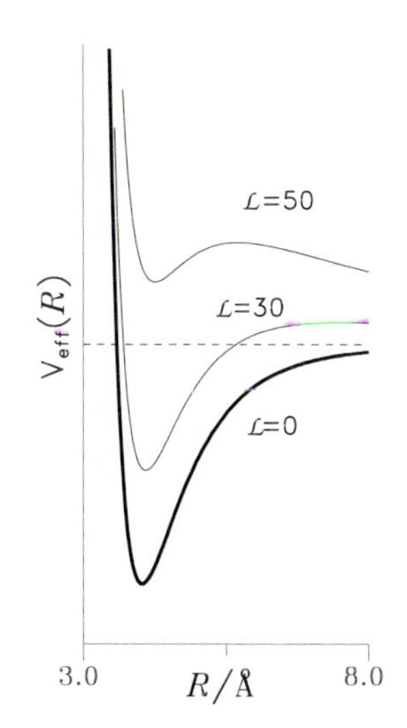

Fig. 3.2 Effective potential energy curve for two atoms.

possesses a barrier, known as the *centrifugal barrier*, which increases in magnitude with increasing orbital angular momentum.

In general, the equations of motion must be integrated numerically, subject to a set of initial conditions. These are the initial relative kinetic energy of the atoms, the impact parameter (which also serves to define the orbital angular momentum via eqn 3.3), the initial separation, R, and the direction of the orbital angular momentum (this defines the plane of the collision). The resulting 'trajectory' describes the time dependence of the position, R, and angle, ϑ, of a particle with an effective mass, μ, experiencing a potential energy $V(R)$.

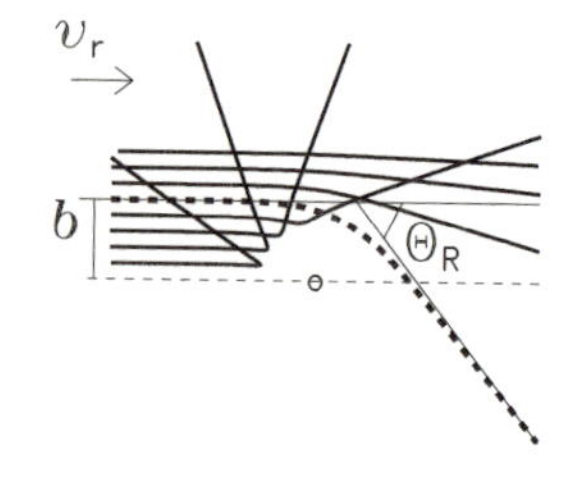

Fig. 3.3 The classical collision between two structureless atoms as a function of impact parameter.

Figure 3.3 illustrates a series of classical trajectories for a fixed initial kinetic energy, but different values of the impact parameter, b (and hence orbital angular momentum, L). The potential energy curve used in these calculations is the Lennard–Jones potential, eqn 2.3. For collisions occurring at the largest b, the impinging atom is barely deflected by the potential energy at all, and its motion is almost unperturbed by the presence of the other atom. These collisions are characterized by small scattering angles, $\theta \sim 0$, as shown in Fig. 3.3; scattering at angles $\theta < 90°$ is usually classified as the *forward* scattering. As the impact parameter is reduced, the atoms experience first the long range attractive part of the potential energy curve, and are pulled towards each other during collision, leading to scattering angles $\theta > 0$. However, as b is decreased still further, the atoms start to experience the repulsive, short range region of the potential, and the scattering begins to move forward once more. Finally, for the smallest impact parameters, the scattered atoms emerge in the *backward* hemisphere, $\theta > 90°$. When the collision encounter is head-on, $b = 0$, then the atoms are scattered directly backward at $\theta = 180°$.

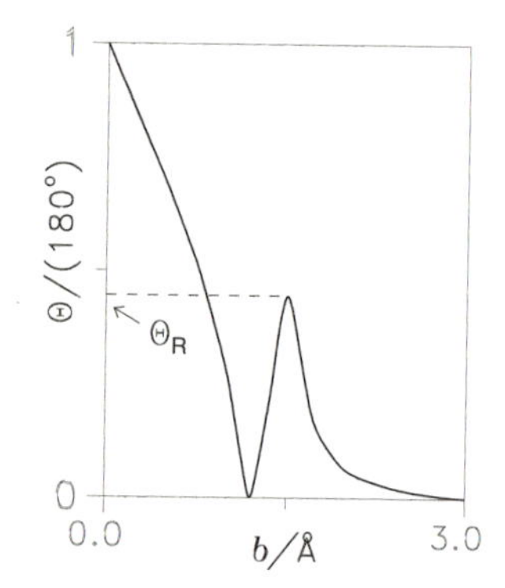

Fig. 3.4 Plot of the scattering angle versus impact parameter for atom–atom scattering.

The trajectories illustrated in Fig. 3.3 may be represented in an alternative fashion, by plotting the scattering angle, θ, against the impact parameter of the collision, once more at fixed collision energy, ϵ_t (see Fig. 3.4). Collision at a given impact parameter is seen to lead to scattering at a specific angle, θ. This is a unique feature of atom–atom scattering, and is a consequence of the spherical, structureless nature of the colliding particles; the direct relationship between impact parameter and scattering angle is not present in reactive collisions. The *rainbow* scattering angle, θ_R, is also defined in Figs 3.3 and 3.4. This occurs at the impact parameter at which the gradient of the line in Fig. 3.4 is zero, $|d\theta/db| = 0$. θ_R depends on the initial kinetic energy, and occurs at an impact parameter at which the *attractive*, long range, part of the potential energy curve gives rise to the largest deflection. Note that scattering into specific angles less than the rainbow angle can arise from collisions at as many as three different values of the impact parameter.

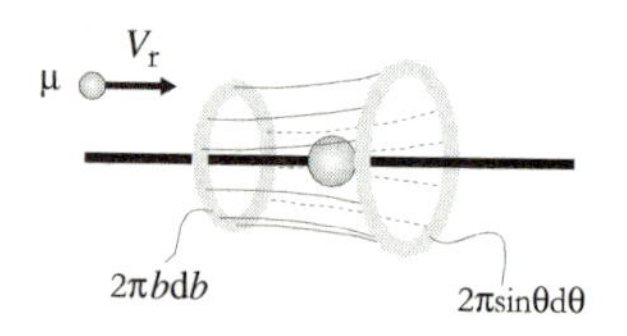

Fig. 3.5 Collisions in the impact parameter range $b \to b + db$ lead to scattering into angles in the range $\theta \to \theta + d\theta$.

It is possible to express the differential cross-section for elastic scattering directly in terms of the impact parameter of the collision. This can be achieved by noting that the probability, $d\wp$, of a collision at impact parameters in the range b to $b + db$ may be written

$$d\wp = \frac{1}{\sigma} 2\pi b \, db, \tag{3.5}$$

where σ is the integral *collision* cross-section. As illustrated in Fig. 3.5, collisions over this range of impact parameters lead to scattering into well-defined angles, in the range θ to $\theta + d\theta$, and thus we may write the probability of scattering into these angles

$$\mathrm{d}\wp = \frac{2\pi}{\sigma}\frac{\mathrm{d}\sigma}{\mathrm{d}\omega}\sin\theta\,\mathrm{d}\theta. \tag{3.6}$$

Given the direct relationship between impact parameter and scattering angle, these two probabilities must be equal and, on rearranging the resulting expression, we obtain the following equation for the differential collision cross-section:

$$\frac{\mathrm{d}\sigma}{\mathrm{d}\omega} = \frac{b}{\sin\theta\left|\frac{\mathrm{d}\theta}{\mathrm{d}b}\right|}. \tag{3.7}$$

The differential cross-section is only a function of θ, and the scattering process has azimuthal, or cylindrical, symmetry about the initial relative velocity vector, $\mathbf{v}_r$, which reflects the random distribution of impact parameters. Reactive collisions also possess this symmetry, provided the reactant molecules are not oriented by the experimenter. Thus, the 2π factor in the equation arises from integration over azimuthal angles ϕ (see Fig. 3.5).

The scattering of hard spheres of diameter d provides a relatively simple illustration of the use of this equation. Firstly, an equation linking the scattering angle to impact parameter is required: this may be obtained from simple trigonometry, with the help of Fig. 3.6, which yields

$$\cos\tfrac{\theta}{2} = \frac{b}{d}.$$

Differentiating this equation with respect to b,

$$\left|\frac{\mathrm{d}\cos\frac{\theta}{2}}{\mathrm{d}b}\right| = \tfrac{1}{2}\sin\tfrac{\theta}{2}\left|\frac{\mathrm{d}\theta}{\mathrm{d}b}\right| = \frac{1}{d},$$

followed by substitution into eqn 3.7, gives the differential cross-section

$$\frac{\mathrm{d}\sigma}{\mathrm{d}\omega} = \frac{\frac{1}{2}\sin\frac{\theta}{2}\cos\frac{\theta}{2}}{\sin\theta}d^2 = \frac{d^2}{4}.$$

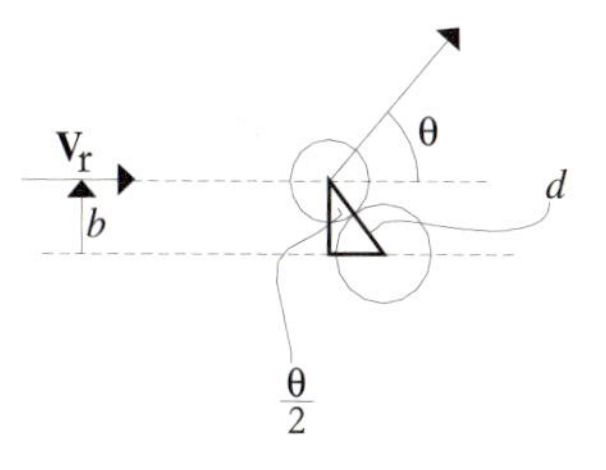

Fig. 3.6 Relationship between b and θ for hard sphere collision.

The differential cross-section for the scattering of two hard spheres is thus independent of scattering angle; i.e. it is *isotropic*. The integral, hard sphere collision cross-section can be obtained by integrating the differential cross-section over all scattering angles, which leads to

$$\sigma = \int_0^{2\pi}\int_0^{\pi}\frac{\mathrm{d}\sigma}{\mathrm{d}\omega}\sin\theta\mathrm{d}\theta\,\mathrm{d}\phi = \pi d^2.$$

Eqn 3.7 has some striking, even alarming, properties when applied to the scattering of atoms rather than hard spheres. It suggests that the classical differential cross-section tends to infinity when $\sin\theta$ tends to zero (e.g. as $\theta \to 0°$), and when $|\mathrm{d}\theta/\mathrm{d}b|$ tends to zero, which occurs at the rainbow angle. The differential cross-section for the scattering off a Lennard–Jones potential is shown in Fig. 3.7. It is seen to be undefined at low scattering angles, and at the rainbow angle. The low scattering intensity in the backwards direction reflects the low probability of head-on collisions (as on a dartboard, the 'bull's-eye' represents a small target area). The fact that there is zero probability of a collision occurring with $b = 0$ ensures that the differential cross-section does not tend to infinity at $\theta = 180°$ (see eqn 3.5).

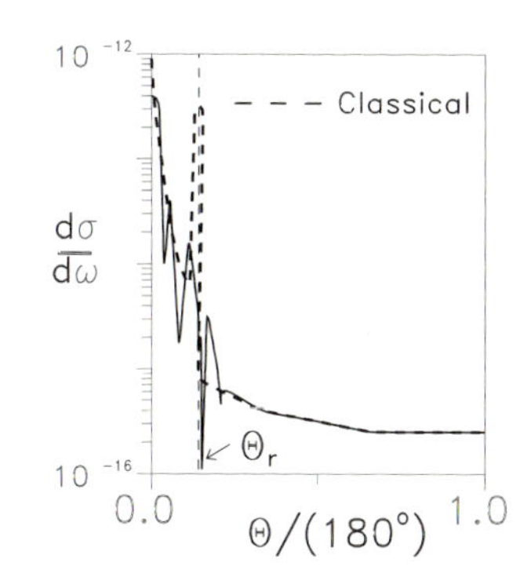

Fig. 3.7 Illustration of the quantum and classical mechanical differential cross-sections for atom–atom scattering.

The classical differential collision cross-section tends to infinity at small scattering angles for any potential which does not equal zero beyond some finite separation, d, (such as the hard sphere potential), since, unless the potential is exactly zero, some finite deflection will occur even at very large impact parameters. Small deflection, large b collisions make a significant contribution to the classical differential collision cross-section, because such collisions have a high probability of occurring, as reflected by the b weighting in eqn 3.7.

Quantum elastic scattering of two atoms

The rigorous treatment of atom–atom scattering uses quantum mechanics. The problem of infinities in the forward direction and at the rainbow angle, found in the classical differential cross-section, are not encountered using quantum mechanics. The reason for this can be understood with the help of Heisenberg's uncertainty principle. In classical mechanics, large orbital angular momentum collisions lead to small, but (in principle) measurable deflections. For quantum mechanical particles, the uncertainty principle places a lower bound on the precision with which the position and momentum of a particle can be simultaneously determined. This, in turn, limits the minimum observable deflection angle, since precise measurement of the deflection angle would allow precise determination of the position and momentum of the target atom.

An example of a quantum mechanically calculated differential cross-section is illustrated in Fig. 3.7, where the data are compared with the classical results. In contrast with the latter, the quantum mechanical differential cross-section is finite at all scattering angles. It also displays a highly oscillatory structure in the region of the classical rainbow. Classically, there are three distinct trajectories which lead to scattering at angles below the rainbow angle (see Fig. 3.4), and we may think of the interference between quantum mechanical waves along these three paths as leading to the pattern shown in Fig. 3.7.

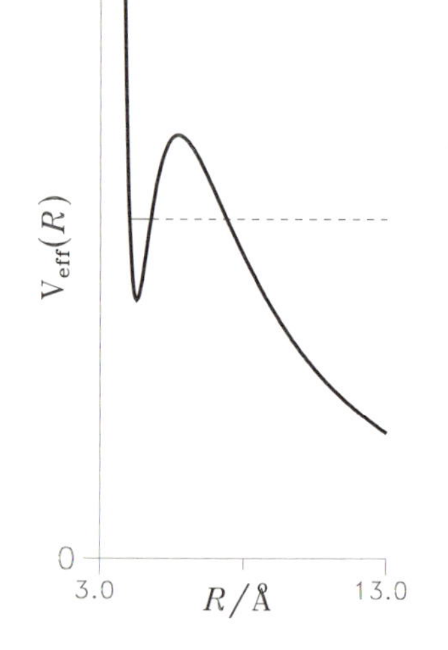

Fig. 3.8 A quasi-bound state on the effective potential, which is responsible for resonance scattering.

Another feature of quantum elastic scattering is *resonance scattering*. Classically, the radial motion of the atoms was shown to occur on an effective potential, determined by the sum of the true potential energy curve and the centrifugal kinetic energy. At certain energies and orbital angular momenta, the effective potential may support *quasi-bound* quantum states; i.e. metastable states which can decay via tunnelling through the effective potential (see Fig. 3.8). If the kinetic energy of the scattering atoms is close to that of one of the states, a large increase in collision cross-section, known as a *shape resonance*, is observed, the width of which is determined by the tunnelling probability through the centrifugal barrier. The angular distribution of the scattered atoms also changes at energies close to a resonance, reflecting the fact that the scattering atoms have a finite probability of existing within the well on the effective potential, where they orbit about each other until they tunnel out again.

3.2 Reactive scattering

The opacity model for direct reactions

In reactive scattering, there is no direct relationship between impact parameter and scattering angle. This is because the impact parameter and kinetic energy of a reactive collision are insufficient initial conditions to define the reactive scattering event. For each reactive collision at given impact parameter and collision energy, the reactants will be oriented in different ways, and may possess different amounts of internal excitation. Thus, there will be a distribution of scattering angles generated from collisions at single values of b and ϵ_t.

However, in direct reactive collisions, some correlation between impact parameter and scattering angle may be preserved. In the *opacity model* of chemical reactions, head-on collisions are assumed to favour backward scattering of the reaction products, while glancing collisions are supposed preferentially to produce forward scattered products, just as in the elastic scattering case described above. Within this model, the differential cross-section obtained for elastic scattering, eqn 3.7, is rewritten

$$\frac{\mathrm{d}\sigma_r}{\mathrm{d}\omega} = \frac{P(b)\,b}{\sin\theta\left|\frac{\mathrm{d}\theta}{\mathrm{d}b}\right|} \qquad b \le b_{\max}, \tag{3.8}$$

where the opacity function, $P(b)$, is introduced to allow for the impact parameter dependence of reaction probability. The reaction cross-section may be written either in terms of an integral of the differential cross-section over scattering angles (eqn 1.5), or of the opacity function over impact parameters (eqn 1.4).

Assume that the opacity function takes the simple form $P(b) = 1$ for $b \le b_{\max}$, and zero otherwise. Reactions which proceed preferentially via head-on collisions could be modelled by choosing a low value of $b_{\max}$. Eqn 3.8 suggests that such reactions will display backward scattering in the differential cross-section, because the collisions which might be expected to lead to forward scattering, those at large b, are unreactive (i.e. have $P(b) = 0$). Furthermore, reactions yielding backward scattered products are predicted also to have small reaction cross-sections.

In contrast, eqn 3.8 suggests that reactions which take place over a much larger range of impact parameters, and thus possess larger values of $b_{\max}$, should yield forward scattered reaction products, analogous to the behaviour observed for atom–atom collisions. These reactions are predicted also to have large reaction cross-sections.

One reason for the predicted forward scattering is that the flat opacity function is weighted by the impact parameter b (see eqn 3.8). The presence of this factor reflects the higher probability of high impact parameter collisions: it is just such collisions which lead to forward scattering.

The opacity model provides a qualitative guide to the interpretation of differential cross-sections for direct chemical reactions—i.e. those reactions which do not proceed via the formation of intermediate complexes (see Section 3.3). In the following section, we describe in detail how differential cross-sections can be obtained more quantitatively, from classical trajectory calculations.[13]

The quasi-classical trajectory method

The material contained in this section is more advanced and could be omitted at first reading.

Figure 3.9 illustrates the coordinate system commonly employed in three-atom reaction dynamics calculations. We may extend the classical Hamiltonian describing the relative motion of atom–atom elastic scattering, eqn 2.6, to that appropriate for three-atom reactions by including a term to allow for the kinetic energy associated with the internal rotational and vibrational motion of the diatomic, $\mathbf{p}^2/2\mu_{BC}$, where μ_{BC} is the reduced mass of the diatomic molecule; i.e.

$$H = \frac{\mathbf{P}^2}{2\mu} + \frac{\mathbf{p}^2}{2\mu_{BC}} + V(R, r, \gamma), \tag{3.9}$$

where

$$\frac{\mathbf{P}^2}{2\mu} \equiv \frac{\mathbf{P}\cdot\mathbf{P}}{2\mu} = \frac{1}{2\mu}\left(P_X^2 + P_Y^2 + P_Z^2\right)$$

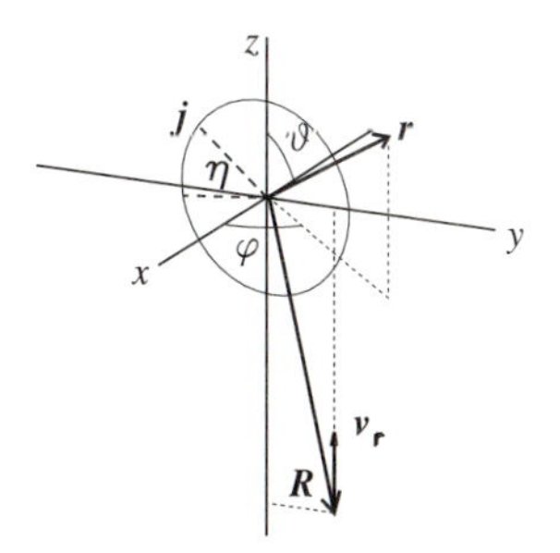

Fig. 3.9 The coordinate system employed for three-atom reactions.

and

$$\frac{\mathbf{p}^2}{2\mu_{BC}} \equiv \frac{\mathbf{p}\cdot\mathbf{p}}{2\mu_{BC}} = \frac{1}{2\mu_{BC}}\left(p_x^2 + p_y^2 + p_z^2\right)$$

are the kinetic energies associated with the relative motion of A and BC and with the internal motion of the BC diatomic respectively. (μ, in these expressions, is the reduced mass of the A + BC reactant pair, i.e. $m_A m_{BC}/M$, where M is the total reactant mass.) Analogous to atom-atom scattering, the kinetic energy terms in eqn 3.9 can be recast in terms of the radial and orbital (centrifugal) kinetic energies,

$$\frac{\mathbf{P}^2}{2\mu} = \tfrac{1}{2}\mu\dot{R}^2 + \frac{L^2}{2\mu R^2},$$

and the vibrational and rotational kinetic energies of the diatomic reactant,

$$\frac{\mathbf{p}^2}{2\mu_{BC}} = \tfrac{1}{2}\mu_{BC}\dot{r}^2 + \frac{j^2}{2\mu_{BC} r^2},$$

where $\mathbf{j}$ is the angular momentum of the diatomic reactant of magnitude, j.

For the X and x components of $\mathbf{R}$ and $\mathbf{r}$, and the associated momenta, Hamilton's equations read

$$\frac{dR_X}{dt} = \frac{P_X}{\mu}$$
$$\frac{dP_X}{dt} = -\frac{\partial V}{\partial R_X}$$
$$\frac{dr_x}{dt} = \frac{p_x}{\mu_{BC}}$$
$$\frac{dp_x}{dt} = -\frac{\partial V}{\partial r_x}.$$

Analogous expressions hold for the remaining components of $\mathbf{R}$ and $\mathbf{r}$, leading to 12 coupled differential equations in total.

Because the scattering event has cylindrical symmetry about $\mathbf{v}_r$, it is not necessary to run trajectories explicitly with different X and Y position and momentum components, provided the orientation of the diatomic is randomly selected (see below).

Trajectories are generated by solving Hamilton's coupled, simultaneous differential equations of motion, subject to a specified set of initial conditions. For each trajectory, the latter are sampled randomly (in so-called *Monte Carlo* fashion) in the following way. With reference to Fig. 3.9, the initial relative position and momentum of the reactants may be written

$$(R_X, R_Y, R_Z) = \left(0, b, -\sqrt{R_i^2 - b^2}\right),$$

and

$$(P_X, P_Y, P_Z) = (0, 0, \mu v_r),$$

where R_i is the initial radial separation, chosen so that the interaction potential energy between A and BC is negligible, and v_r is the initial relative speed. The choice of b, the impact parameter, depends on the property to be evaluated. If opacity functions ($P(b)$) are required, then b is usually selected randomly in the interval $0 \to b_{max}$, where the latter is chosen so that trajectories run with $b \geq b_{max}$ do not lead to reaction. If, on the other hand, reaction cross-sections or differential cross-sections are required, it is more efficient to sample b^2 randomly between 0 and b_{max}^2. The effect of this is to weight the sampling of trajectories to high b values, consistent with the fact that the probability of a collision in the range $b \to b + db$ is $2\pi b\, db$.

The Cartesian components of the initial position vector of the diatomic, $\mathbf{r}$, the centre-of-mass of which represents the origin of Fig. 3.9, may be expressed

$$r_x = r \sin\vartheta \cos\varphi$$
$$r_y = r \sin\vartheta \sin\varphi$$
$$r_z = r \cos\vartheta,$$

where φ and $\cos\vartheta$ are chosen randomly in the intervals $0 \to 2\pi$ and $-1 \to +1$, respectively. The selection of the initial momentum associated with the diatomic molecule is simplified if the trajectories are started at initial bond lengths, $r \equiv R_{BC}$, set equal to a turning point in the classical vibrational motion, which for a harmonic oscillator would be

$$r_{\pm} = r_e \pm \sqrt{\frac{2h\nu}{k}(v + \tfrac{1}{2})},$$

where ν is the vibrational frequency, k is the force constant, and v is the initial vibrational state of BC. To ensure random selection of the vibrational phase of the diatomic molecule, the initial separation of the reactants, R_i, is modified by adding an amount $\Delta R_i = \frac{1}{2} v_r \tau_{vib} \kappa$, where κ is a random number between 0 and 1, and $\tau_{vib} = 1/\nu$ is the vibrational period. This means that by the time the reactants reach separation, R_i, the vibrational phase of BC will be random. It also means that the initial momentum associated with the vibrational motion of BC, $\mu_{BC}\dot{r}$, is zero, because initially (i.e. at $R = R_i + \Delta R_i$) the BC molecule is at a classical turning point, where the total vibrational energy of BC is in the form of potential energy.

In the QCT method the initial 'states' of the diatomic are chosen to correspond to the known quantum mechanical energy levels of the molecule (see Chapter 2).

The only remaining initial conditions to specify are those defining the orientation of the initial rotational angular momentum of BC. For a diatomic molecule, the angular momentum must be (randomly) oriented in a plane perpendicular to the bond axis, **r**. In terms of the magnitude of the initial angular momentum of BC, j $(\equiv \hbar\sqrt{j(j+1)})$, and the orientation angle η defined in Fig. 3.9, the components of the linear momentum of BC may be written

$$p_x = \frac{j_x}{r} = -\frac{j}{r}(\sin\varphi\cos\eta + \cos\varphi\cos\vartheta\sin\eta)$$

$$p_y = \frac{j_y}{r} = \frac{j}{r}(\cos\varphi\cos\eta - \sin\varphi\cos\vartheta\sin\eta)$$

$$p_z = \frac{j_z}{r} = \frac{j}{r}(\sin\vartheta\sin\eta),$$

where η is chosen randomly in the interval $0 \to 2\pi$.

Once the initial conditions have been selected, the equations of motion are integrated numerically until the atoms are widely separated once more. Reaction to produce AB + C, for example, is deemed to have occurred if the bond lengths R_{AC} and R_{BC} are both larger than a threshold value at which the influence of the potential energy surface in these coordinates is negligible. Similar criteria define both non-reactive (elastic and inelastic) collisions and reaction to produce AC + B. Each trajectory must then be analysed to establish the rotational and vibrational energies and states of the AB product and the scattering angle. To this end it is helpful to express the positions and momenta of the atoms in a new set of (CM) coordinates, $\mathbf{R}'$ and $\mathbf{r}'$, appropriate for the AB and C products of reaction, as illustrated in Fig. 3.10. Assignment of the products to 'quantum states' is achieved using an energy (and angular momentum) binning procedure, in which the central energy of the bin corresponds to a known energy level of the product. A similar approach is taken to determine the differential cross-section. The scattering angle (defined conventionally as the angle between the CM velocity vectors of the incoming atom and the outgoing AB diatomic) for a given trajectory is given by $\cos\theta = -\dot{R}'_Z/\dot{R}'$, where $\dot{R}'$ is the relative velocity of the products. The latter must be expressed in terms of the reactant coordinates, for which $\mathbf{v}_r$ is parallel to Z. After running a number of trajectories, N, employing randomly selected

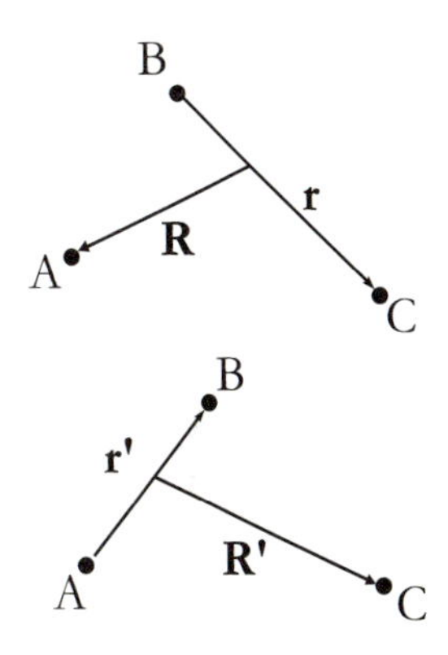

Fig. 3.10 Reactant and product coordinates for three-atom reactions.

The reactant and product (CM) coordinate systems are related by the following transformation:

$$\begin{pmatrix} \mathbf{R}' \\ \mathbf{r}' \end{pmatrix} = \begin{pmatrix} -a & b \\ -c & -d \end{pmatrix} \begin{pmatrix} \mathbf{R} \\ \mathbf{r} \end{pmatrix},$$

where $a = m_A/m_{AB}$, $b = \mu_{BC}/\mu'$, $c = 1$, and $d = m_C/m_{BC}$; μ' is the reduced mass of the product, AB and C, pair. The same transformation applies both to the Cartesian components of each of these position vectors, and to the associated velocities (i.e. the time derivative of the above equation).

initial conditions, the differential cross-section is obtained by binning the trajectories in $\cos\theta$ 'bins', of width $\Delta\cos\theta$, and evaluating the equation

We have assumed that the *square* of the impact parameter has been sampled uniformly, rather than b itself, as described above. This procedure circumvents the need to integrate explicitly over the opacity function when calculating the differential and reaction cross-sections.

$$\frac{d\sigma_r}{d\omega} = \frac{N_r(\theta)}{N}\frac{\pi b_{max}^2}{2\pi\Delta\cos\theta} \qquad N \to \infty,$$

where $N_r(\theta)$ is the number of *reactive* trajectories in the range $\cos\theta \to \cos\theta + \Delta\cos\theta$.

Molecular beam and other scattering experiments

For a chemical reaction occurring at a well-defined initial kinetic energy, for example

$$A + BC \longrightarrow AB(v', j') + C,$$

the reaction products may be born in a variety of internal (v', j') states. They will be endowed, therefore, with a distribution of relative speeds, reflecting the population of these different product internal states. The ideal experiment would be to measure the intensity or flux of the scattered products as functions of laboratory scattering angle *and* speed, for a well-defined reactant collision energy and BC internal state. Attempts to achieve this goal most commonly employ the crossed molecular beam technique, coupled with the universal detector, the (time-of-flight) electron impact ionization mass spectrometer. A new generation of laser-based spectroscopic detection techniques is also occasionally used in the crossed molecular beam environment. Although these are less universally applicable than mass spectrometry, they offer the advantage of providing higher product internal energy resolution than more conventional methods.

A detailed description of the molecular beam experiment can be found in Bernstein's *Chemical dynamics via molecular beam and laser techniques*.

The crossed molecular beam experiment. The generation of molecular beams is generally achieved in one of two ways, depending on the reactants to be employed. Many of the early applications of the molecular beam strategy were to reactions involving alkali metal atoms. In these experiments, an *effusive* beam of such atoms is generated by heating the metal in an oven and allowing the gas-phase atoms to escape via a small pin-hole into a high vacuum scattering chamber. The molecular beam thus generated is characterized by a rather broad speed distribution, characteristic of the temperature of the oven source.

Molecular beams of stable molecular reactants are generally produced by an alternative method. This involves adiabatic expansion of very high pressures of the relevant gas, often 'seeded' in an inert carrier gas such as He, through a pin-hole or *nozzle*. Depending on the nozzle diameter and backing pressure employed, a directed molecular flow can be generated, which is characterized by very high molecular speeds in the beam direction, and very low speeds orthogonal to it. Pulsed molecular beams may also be produced by opening the nozzle in short bursts using, for example, a piezoelectric device. Although such procedures reduce the duty cycle of the experiment (the fraction of time during which useful data can be collected), pulsing the molecular beam reduces the total gas flow through the apparatus, allowing relatively small vacuum pumps to be employed. It also enables experiments to be conducted with high reactant backing pressures, which produces cold and intense molecular beams. Pulsed molecular beams are ideally suited to applications in which the detection technique is also pulsed, as is often the case with laser-based methods. Both the pulsed and continuous, non-effusive methods have the additional advantage that the rotational (and, to a lesser extent, vibrational) degrees of freedom of the reactant molecules are cooled by collisional relaxation in the immediate post-nozzle region: rotational temperatures of a few kelvin are routinely achieved. More recent developments have focused primarily on extending the technique to allow the generation of beams of atomic and radical species, for example by chemical reaction or electrical discharge immediately prior to the molecular expansion.

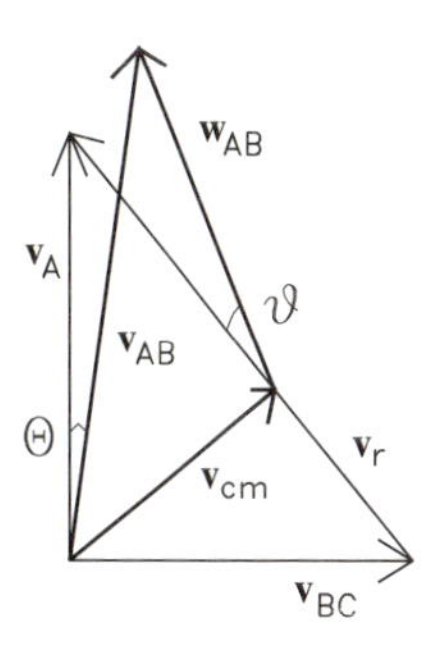

Fig. 3.11 Newton diagram illustrating the laboratory to CM transformation.

The molecular beams, generated using the methods just described, are usually collimated using a *skimmer* prior to crossing in the interaction region of the scattering chamber. The latter is maintained at very low pressures to avoid secondary collisions, which would lead to energy and momentum transfer and distort the angular and product state information of interest. The beam velocities and hence the collision energy may be tuned by employing different backing pressures and carrier gases, or by allowing the reactant beams to cross at different intersection angles. The reaction products are detected at some well-defined distance from the interaction region and at a variable laboratory scattering angle using electron impact, time-of-flight mass spectrometry.

The time-of-flight mass spectrometer is a variant of conventional mass spectrometry which allows the number of products arriving at the detector to be resolved as a function of time. Knowledge of the path length from the scattering centre allows the speed of the products to be calculated from the measured arrival times. The timing of the experiment is usually controlled by chopping or pulsing the reactant molecular beam: the time resolution allows both the mass and the speed distribution of the products to be determined, as a function of laboratory scattering angle.

The raw data from these experiments consist of the intensity of the reaction product (proportional to the product number density) as a function of the laboratory (LAB) speed and laboratory angle, usually defined with respect to the direction of one of the reactant molecular beams. To obtain the differential cross-section, and the relative speed of the reaction products, the signal intensity must be converted to flux (see eqn 3.1) by multiplying by the product speed, and then transformed from the laboratory frame to the centre-of-mass (CM) frame. The *Newton diagram* shown in Fig. 3.11 illustrates the key features of this transformation, where '**v**' is used to denote a LAB velocity and '**w**' to denote a velocity in the CM frame. The information obtained is usually plotted in the form of a product flux velocity–angle contour plot, in which the CM corresponds to a point at the centre of the diagram, and contours represent loci of equal product flux as a function of CM scattering angle and product speed. Such a diagram is illustrated in Fig. 3.12.[17]

Laser-based techniques. Although crossed molecular beam experiments provide well-defined reactant kinetic energies and good angular resolution, one disadvantage of the electron impact ionization mass spectrometric detector is its comparatively low product kinetic energy resolution. This low resolution means that, except in very favourable cases, it is not possible to use the technique to obtain differential cross-sections with product quantum state resolution.

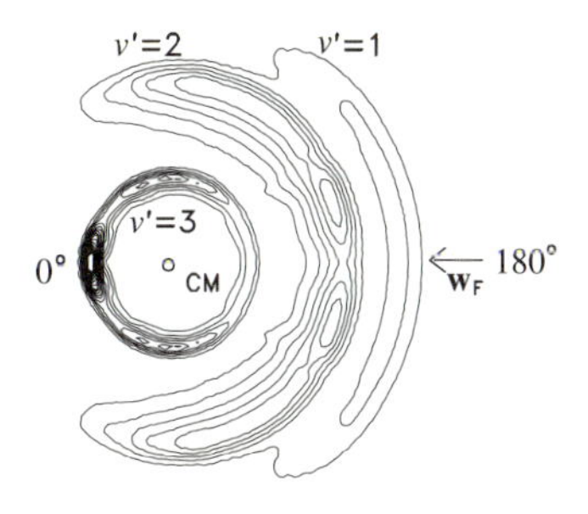

Fig. 3.12 HF product flux contour plot for the reaction $F + H_2$.

Spectroscopic techniques for probing the reaction products have traditionally been applied to determine the population distribution over product internal states (see Section 4.1). Such techniques, which generally employ highly monochromatic laser radiation, inherently offer greater product state resolution, but are rarely used in conjunction with the crossed molecular beam environment, because the number density of product molecules born in specific quantum states is typically too small to allow reliable detection. One pioneering study, which did combine the crossed molecular beam with laser-based detection methods, employed the laser-induced fluorescence technique to probe the OH products of the H + NO_2 reaction.[14] The study made use of the Doppler effect to obtain product speed and angular distributions for rovibrationally state-selected OH products of the reaction. Continued improvements in detection sensitivity are beginning to result in more widespread use of techniques such as this to obtain both product internal state distributions and state resolved differential cross-sections.[15]

A more recent alternative strategy,[16] which does not employ crossed molecular beams, uses pulsed laser photolysis of a suitable precursor molecule to generate 'beams' of fast-moving atom or radical species at low pressures (see Section 4.1). Pulsed laser probing of the reaction products after a short 'pump–probe' delay then makes use of the Doppler effect to obtain information on the velocity distribution of the reaction products, from which the CM differential cross-section can be obtained. The technique is ideally suited to reactions involving three atoms, and to obtain the highest energy and angular information the use of translationally cold precursor and 'target' reactant molecules is required. The translational cooling is achieved using the jet-cooling technique (see Section 2.3), by co-expanding the precursor and target molecules in a pulsed molecular beam. The reactant number densities generated by this method are still orders of magnitude higher than in the crossed molecular beam experimental configuration.

3.3 Case studies

Direct reactions

The velocity–angle contour map for the reaction

$$F + H_2 \longrightarrow HF + H$$

was shown in Fig. 3.12.[17] The majority of HF products are found in the backward hemisphere, peaking at scattering angles around 180° relative to the reactant F atom direction. In conjunction with the low reaction cross-section, the data are suggestive of preferred reaction at low impact parameters. The wide rovibrational energy level spacing in HF facilitates the resolution of different product vibrational states in the velocity–angle contour plot (labelled v' in Fig. 3.12). Therefore, plots such as these also provide information on the distribution of HF molecules in different product vibrational states (see Section 4.3). Rather strikingly, HF products born in the highest vibrational levels (corresponding to the innermost circle in Fig. 3.12) are scattered selectively into the forward hemisphere. Classical trajectory calculations[19] on an accurate potential energy surface[4], which account for many of the features displayed experimentally, reveal that the highly vibrationally excited products arise from comparatively high impact parameter collisions. However, one failing of the classical calculations is that the HF($v' = 3$) differential cross-sections do not show sufficient intensity in the forward direction, compared with experiment. Quantum mechanical calculations[20] on the same surface show better agreement with experiment. They display enhanced scattering into the forward hemisphere, which arises from quantum mechanical tunnelling. Classically, reaction at very large impact parameters is limited at a given collision energy by the height of the centrifugal barrier on the effective

This reaction, together with that of D + I_2 described below, provide examples of *kinematically constrained* reactions (see Section 4.3).

More recent experiments[18] have yielded product *rotationally* resolved differential cross-sections.

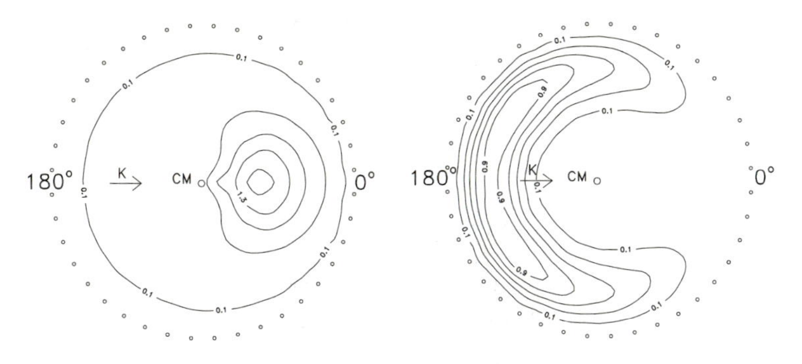

Fig. 3.13 KI product flux contour plots for the reactions $K + I_2$ (left) and $K + CH_3I$ (right).

potential. Quantum mechanically, reaction can occur at larger orbital angular momenta, because the reactants can tunnel through the centrifugal barrier. These highest L reactive collisions are scattered primarily in the forward direction.

Further examples of differential reactive cross-sections are provided by the two reactions[21,22]

$$K + I_2 \longrightarrow KI + I$$

and

$$K + CH_3I \longrightarrow KI + CH_3.$$

Both reactions proceed via the *harpoon mechanism*, but the two display contrasting characteristics associated with preferential reaction at large and small impact parameters. This arises because the charge transfers by which these reactions proceed occur at very different reactant separations in the two cases. For the $K + I_2$ reaction, charge transfer occurs at large R, and the reaction probability, $P(b)$, is close to unity over a very large range of impact parameters. Thus both the integral and the differential reaction cross-sections are dominated by the more probable high impact parameter collisions, which lead to forward scattering (see Fig. 3.13). By contrast, the reactants in the $K + CH_3I$ reaction must be very close together before charge transfer can take place. In consequence, the opacity function peaks at small impact parameters, as is evident from the low reaction cross-section and the strong backward peaking differential cross-section, shown in Fig. 3.13.

The velocity–angle maps shown in Fig. 3.13 also provide information on the population of the internal states of the KI products (see Section 4.2).

The angular distribution of the scattered products in the reaction[23]

$$D + I_2 \longrightarrow DI + I$$

provides another, distinctive example (see Fig. 3.14). The DI products are found preferentially in the sideways direction, perpendicular to the reactant relative velocity vector. One interpretation for this scattering is that it reflects a preferential sideways angle of attack of the D atom to the I_2 bond. Because the D atom is much lighter than iodine, the direction of recoil of the DI product is determined by the angle of attack, and is little altered by the momentum carried by the incoming D atom.

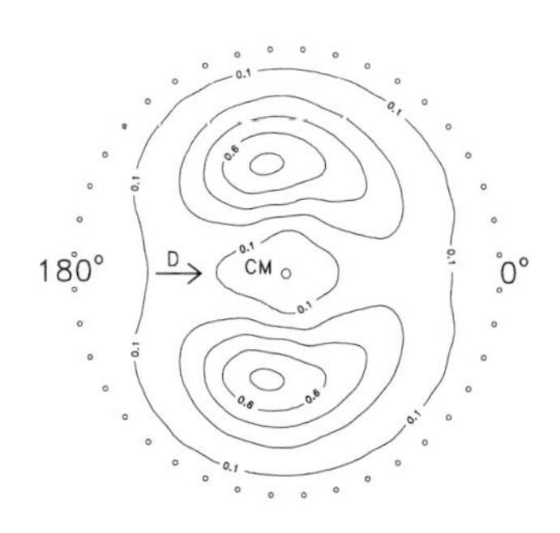

Fig. 3.14 DI product flux contour plot for the reaction $D + I_2$.

As a final example, we turn to an isotopic variant of the simplest of all chemical reactions involving neutral species, namely

$$H + D_2 \longrightarrow DH + D.$$

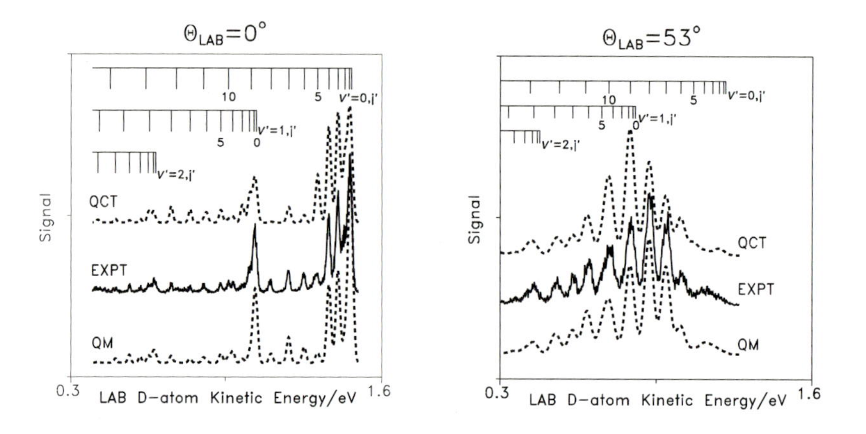

Fig. 3.15 D atom time-of-flight spectra for the reaction $H + D_2$ at the LAB scattering angles shown. These LAB angles correspond approximately to backward (left) and sideways (right) HD scattering. The labelled peaks are associated with the production of HD co-products in different rovibrational states.

The data, shown in Fig. 3.15, were obtained using a state-of-the-art spectroscopic method for ionizing the product D atoms, which allowed a higher product D atom time-of-flight resolution to be obtained than can be had with conventional, electron impact ionization mass-spectrometry.[24] The resulting D atom speed resolution is so high that differential cross-sections of individual rovibrational states of the DH co-product can be resolved. The reaction is dominated by low impact parameter, near-collinear collisions, for which the reaction barrier is a minimum, and the general observation of backward scattering is in qualitative accord with expectation based on the simple opacity model. Collinear collisions at low impact parameters cannot provide a mechanism for generating HD product rotation, and it is significant that little product rotation is observed experimentally. Furthermore, the most sharply backward scattered HD products correspond to those produced in low rotational states. Higher HD rotational states originate from higher impact parameter reactive collisions, and the scattering of these products is less focused in the backward direction. The differential cross-sections and product quantum state populations have been calculated both quantum mechanically and classically. Both calculations, particularly the former, are in good accord with experiment.[24]

Reactions proceeding via a long-lived complex

For reactions which proceed via the intermediacy of a long-lived complex, the opacity model can no longer guide interpretation of the differential cross-section. If the lifetime of the intermediate complex is long compared with the rotational period of the complex, then the approximate correlation between impact parameter and scattering angle observed for direct reactions no longer applies. Such reactions display differential cross-sections which are symmetric about the sideways scattering angle of 90°. The precise shape of the differential cross-section depends on the structure of the complex (e.g. whether it is a prolate or oblate top), and on angular momentum conservation constraints (see Section 4.3).

The rotational period of a complex ($= 2\pi/\omega$) can be estimated classically from the equation $J = I\omega$, where J is the magnitude of the total rotational angular momentum of the complex, I its moment of inertia (about the rotational axis of interest), and ω the angular velocity, in radians s^{-1}. The magnitude of the angular momentum can be estimated quantum mechanically, using the equation $J = \hbar\sqrt{\mathcal{J}(\mathcal{J}+1)}$, where $\mathcal{J}$ is the total angular momentum quantum number of the complex.

Imagine a complex is formed on combination of species A and B with orbital angular momentum L (in this simple example, we assume that the rotational angular momentum of the reactants is zero, and that the orbital angular momentum is conserved). The reactants can approach at any azimuthal angle ϕ (see Fig. 3.5), such that $\mathbf{L}$, which must be perpendicular to $\mathbf{v}_r$, can point anywhere on the circle, as illustrated in Fig. 3.16. For a given direction of $\mathbf{L}$, the collision is confined to a plane, and thus, if the complex survives for many rotational periods, the probability of scattering at angles between θ and $\theta + \mathrm{d}\theta$ must be uniform. Averaging over all collision encounters, with different orientations of $\mathbf{L}$, yields a scattering probability in the ranges $\theta \rightarrow \theta + \mathrm{d}\theta$ and $\phi \rightarrow \phi + \mathrm{d}\phi$ which is independent of polar angles. Thus we may write

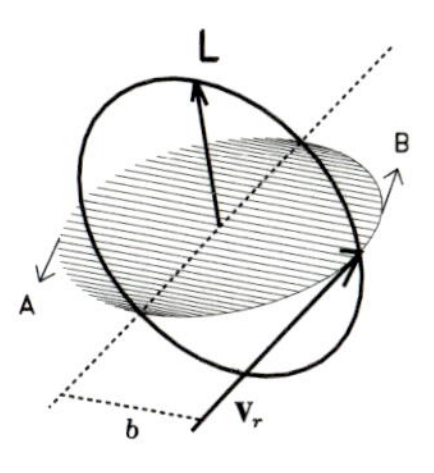

Fig. 3.16 The origin of forward–backward symmetry in the differential cross-section for a long-lived diatomic complex.

$$P(\theta, \phi) \sin \theta \mathrm{d}\theta \mathrm{d}\phi \equiv \frac{1}{\sigma} \frac{\mathrm{d}\sigma}{\mathrm{d}\omega} \sin \theta \mathrm{d}\theta \mathrm{d}\phi = N \mathrm{d}\theta \mathrm{d}\phi,$$

where N is a normalization constant. Rearranging this expression, together with evaluating the constant, yields

$$\frac{1}{\sigma} \frac{\mathrm{d}\sigma}{\mathrm{d}\omega} = \frac{1}{2\pi^2 \sin \theta}.$$

The differential cross-section is predicted to peak symmetrically in the forward and backward directions. An example of a reaction showing a differential cross-section with near-symmetric forward and backward peaks, O + Br_2, is shown in Fig. 3.17.[25]

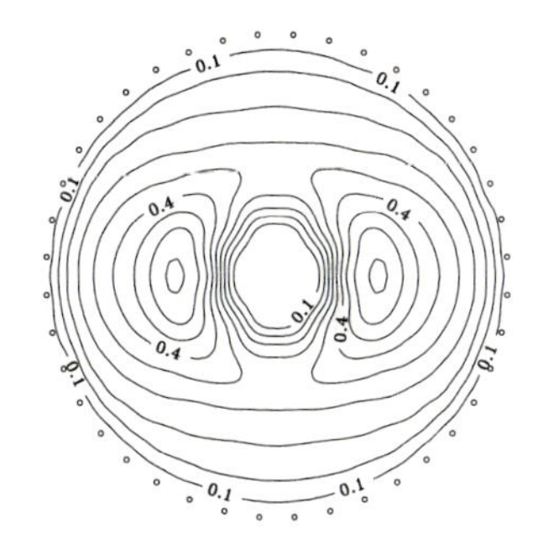

Fig. 3.17 BrO product flux contour plot for the reaction O + Br_2.

More commonly, the lifetime of the reactive intermediate is not long compared with its rotational period (and L is not conserved), and in such circumstances, some asymmetry in the differential cross-section is observed. An example is the insertion reaction of $O(^1D)$ + H_2, which shows a slight preference for backward scattering in the state averaged differential cross-section at low collision energies. The product state resolved differential cross-section for OH($v' = 0, j' = 14$) shows more pronounced backward scattering[26] (see Fig. 3.18), and QCT calculations suggest that this behaviour arises because a small subset of intermediate complexes (in particular, those H_2O complexes which decay to OH($v' = 0$) products in low rotational levels) decay on timescales shorter than the rotational period. At higher collision energies, backward scattering becomes more dominant.[27] This behaviour has been ascribed (in contrast to the above mechanism) to competing reaction on an electronically excited potential energy surface, which does not correlate with the ground electronic state of the water molecule.[5,27] Reaction on this upper surface is believed to occur via a direct abstractive, rather than an insertion mechanism, and yields backward scattered products.

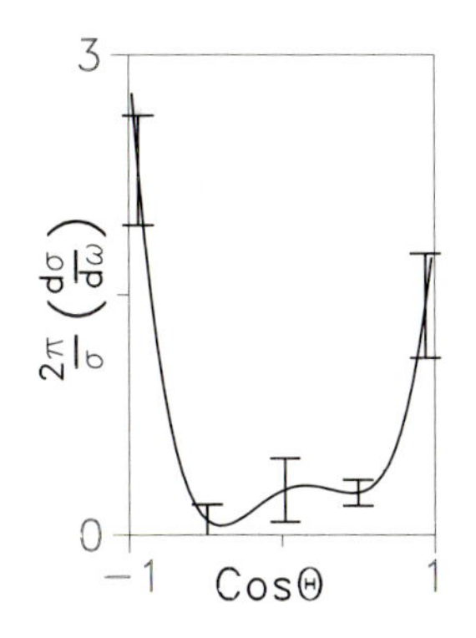

Fig. 3.18 Differential cross-section for the reaction $O(^1D) + H_2$. The detected OH products were in $v' = 0, j' = 14$.

3.4 Stereochemistry

We have seen that differential cross-sections for direct reactions provide some insight into the distribution of impact parameters and angles of attack which lead to reaction. *Stereochemistry* is the area of reaction dynamics concerned with the reactant orientation dependence of chemical reactivity. To obtain more direct stereochemical information it is necessary to pre-orient the reactants, and thereby exert some control over whether collisions take place in preferred collinear ('heads' or 'tails') or side-on configurations.

Stereochemistry or *stereodynamics* is also concerned with the polarization of the angular momentum of the reaction products, i.e. with the angular distribution of the product rotational angular momentum. Polarized light can be employed both to polarize the angular momentum of the reactants and interrogate the angular momentum polarization of the products.[16, 28]

A number of ways have been developed for pre-orienting molecules. The simplest method, which is suitable for reactant molecules which have a dipole moment, is to orient them in a strong electric field. Although the technique can be applied to a wide variety of molecular systems, the main disadvantage of the technique is that the net orientation achieved in the electric field depends on the initial angular momentum of the molecule, in the field-free region of the molecular beam expansion. Although highly oriented molecules can be generated if, initially, only low angular momentum states of the molecule are populated (such as would be the case in the coldest of molecular beams), higher angular momentum states are more difficult to orient, as they undergo wide-angle pendular-type motion in the electric field. One way of dealing with this problem is to pre-select the angular momentum state of reactant molecules prior to their entering the homogeneous electric field region. This can be achieved by using an inhomogeneous hexadecapole electrostatic field, for example.

The inhomogeneous electrostatic field is used to focus molecules in specific J, K, and M_J states through slits placed in the molecular beam flight path. The method is most easily applied to prolate symmetric top reactants.

The molecular beam of oriented reactant molecules is crossed with that of the co-reactants at a well-defined angle to the homogeneous electric field. In the example presented in Fig. 3.19, the products (RbI) of the reaction

$$\mathrm{Rb} + \mathrm{CH_3I} \longrightarrow \mathrm{RbI} + \mathrm{CH_3}$$

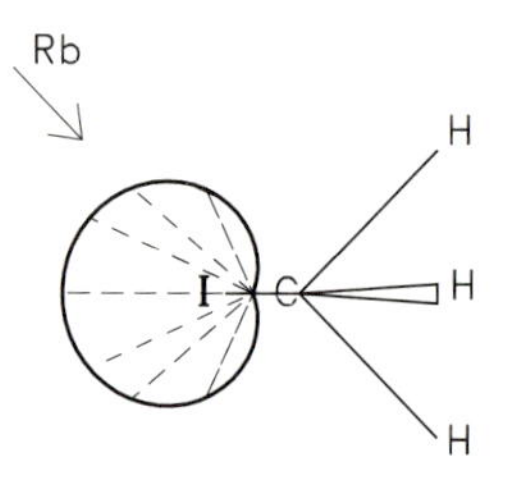

Fig. 3.19 The reaction probability for the Rb + CH_3I reaction as a function of angle of attack between the incoming Rb atom and the C–I bond axis.

were detected mass spectrometrically in the backward scattered direction. The figure represents the reaction probability (for producing backward scattered RbI) as a function of orientational angle between the reactant relative velocity vector and the C–I bond (which coincides with the direction of the dipole moment). The reaction probability peaks when the incoming Rb atoms approach head-on with the iodine end of the methyl iodide molecule, and decays to zero when the Rb attacks tails-on (i.e. towards the CH_3 group of the reactant).[29] Thus, there is a *cone of acceptance* around the iodine end of the molecule, along which the Rb atom must approach if it is to react. The data provide direct experimental evidence for the steric effect, widely invoked in fields such as synthetic organic chemistry.

An alternative technique has been applied to a study of the reaction

$$\mathrm{H} + \mathrm{CO_2} \longrightarrow \mathrm{OH} + \mathrm{CO},$$

which is believed to proceed via the HOCO intermediate.[30] The reaction was initiated by laser photolysis of the HX–CO_2 (X=Cl, Br, and I) van der Waals complex (see Chapter 2):[31] the HBr chromophore absorbs the radiation, generating translationally excited H atoms, whose direction of travel is determined by the geometry of the complex in its zero-point vibrational level. Experiments of this type provide some control over both the impact parameter and orientation angle of the reactants, and have been used to determine energy disposal data, which can be compared with those obtained from the gas phase 'hot atom' reaction (see Chapter 4), and to obtain estimates of the lifetime of the intermediate HOCO complex.

4 State-specific cross-sections

4.1 Experimental considerations

Molecular beam experiments

Molecular beam experiments provide high initial collision energy resolution. Because of this, they are often employed to measure the translational energy dependence of the reaction cross-section (or *excitation function*). The excitation function can (in principle) be obtained by integrating the differential cross-section over scattering angle, but there are more direct methods. For reactions which produce ions, $\sigma_r(\epsilon_t)$ can be obtained by deflecting all the ionic products (irrespective of their scattering angle) into the flight tube of a mass spectrometer, using an electric field. The experiment is therefore sensitive directly to the integral, rather than the differential, cross-section. Alternatively, for reactions with very large cross-sections, such as many harpoon reactions, the excitation function can be obtained from a measurement of the attenuation of the reactant beam flux (i.e. in the forward scattered direction). This simple method is only valid if the majority of collisions, which lead to a deflection in the reactant velocity, also lead to reaction, rather than elastic or inelastic processes.

Information on the disposal of internal energy can also be provided by conventional crossed molecular beam experiments, coupled with mass spectrometric detection. As we have seen, such experiments yield the product flux as a function of both scattering angle and product speed. Hence, from energy and linear momentum conservation, information about the internal energy distribution of the products can be inferred. Examples of higher resolution molecular beam techniques, which rely on laser-based methods of detection, have already been presented (see Section 3.3).

Spectroscopic based experiments

The highest (quantum state) resolution is usually provided by radiation- or laser-based experiments. These methods are generally employed in experiments conducted under 'bulb' rather than crossed molecular beam conditions. Excitation functions have been obtained using spectroscopic methods of product detection. In the flash-photolysis *hot atom* experiment, the atomic reactant is generated photolytically, using pulsed laser radiation. The hydrogen halides are frequently employed as photolysis precursors for translationally excited ('hot') H atoms: the kinetic energy of the H atoms can be varied, either by choice of precursor molecule, or by varying the photolysis wavelength. The product molecules are detected (or 'probed') after a short, known time delay, using, for example, laser-induced fluorescence. To obtain *absolute* (rather than relative) reaction cross-sections, the absolute concentration of the products must be determined. This is often achieved by calibrating the signal intensity against some known standard, such as the signal obtained for the same product species, but generated by photolysis of a precursor molecule with a known absorption cross-section and dissociation quantum yield.

The hydrogen halides are employed partly because they possess absorption continua in a convenient wavelength range, and because the excess energy (the photon energy minus the bond dissociation energy) is almost exclusively channelled into kinetic energy of the light H atoms, as a result of momentum conservation.

Similar calibration procedures are required to convert relative final state resolved reaction cross-sections (i.e. quantum state populations) into absolute product state resolved cross-sections.

If reaction products are born vibrationally or electronically excited then they may be detected by monitoring the (IR or visible) chemiluminescence. The former emission was monitored in many of the pioneering studies of energy disposal in chemical reactions, and IR emission spectroscopy is still widely used today, partly because of its sensitivity to a diverse range of emitting species. Applications of the method have also been facilitated by the emergence of fast Fourier transform techniques. IR chemiluminescence is, however, 'blind' to molecules produced in $v = 0$.

Further details about these and other spectroscopic detection methods can be found in texts given in the Background reading.

Many detection methods rely on high intensity, monochromatic laser radiation. Of these, the most frequently employed are *laser-induced fluorescence* (LIF) and *resonantly enhanced multiphoton ionization* (REMPI). Because ions are produced in the REMPI technique, which can be efficiently detected by mass spectrometry, this method has greater sensitivity than LIF, in which emitted photons are detected. However, the REMPI scheme is a multiphoton process, and it is sometimes more difficult to relate signal intensities to level population than for LIF.

Selective excitation of reactant molecules is more difficult to achieve. The most frequently employed methods involve direct laser absorption (excitation) to specific rovibrational states, but alternative, more sophisticated 'non-linear' optical methods have also been used. A key consideration in all experiments of this type, whether they involve reactant state preparation or product state detection, is the need to avoid collisional relaxation. The problem is most severe in 'bulb' type experiments. To avoid collisional relaxation, low total pressures must be employed, and in laser pump–probe (flash photolysis) experiments, short time-delays between reactant preparation and product interrogation. Because translational and rotational relaxation are generally more efficient than vibrational relaxation, it is somewhat easier to obtain information about the utilization and disposal of vibrational energy than other types of energy.

The time-delays and pressures required to avoid relaxation can be estimated from the collision frequency factor, Z^0_{AB}; see Section 1.3. Experimentalists usually strive to work below *single collision conditions*, which means that the reactants and products suffer less than one collision on the timescale of the experiment.

4.2 Models of energy utilization and disposal

The role of collision energy

Reactions with a barrier. A qualitative appreciation of how the reaction cross-section varies with collision energy is provided by the *line-of-centres* model, which forms the basis of simple collision theory (introduced in Section 1.3). In this model, the reactants are treated as two structureless spherical particles, which react provided the radial kinetic energy is greater than zero at the reaction barrier, $V(R = R_0) = \epsilon_0$; i.e.

See Section 3.1, on which the material presented here is based.

$$\tfrac{1}{2}\mu \dot{R}^2 \geq 0 \qquad \text{at} \qquad R = R_0,$$

where R_0 is the separation of the particles at the barrier. At fixed initial kinetic energy, ϵ_t $(= \frac{1}{2}\mu v_r^2)$, reaction is assumed to occur with unit probability at all impact parameters ($P(b) = 1$) up to some maximum value, b_{max}. The latter is evaluated using the radial equation (eqn 3.4), derived for the collision between two structureless particles. This reads, at the barrier,

$$\tfrac{1}{2}\mu \dot{R}^2 = \epsilon_t - \epsilon_t \frac{b^2}{R_0^2} - \epsilon_0. \qquad (4.1)$$

The second term on the right of this equation is the centrifugal kinetic energy associated with the orbital motion of the reactants. At fixed collision energy, the radial kinetic energy at the barrier is zero when the centrifugal kinetic energy, and, hence, the impact parameter, are at maxima. Thus, setting the right of eqn 4.1 to zero leads to the following equation for b_{max}:

$$b_{max}^2 = R_0^2\left(1 - \frac{\epsilon_0}{\epsilon_t}\right). \tag{4.2}$$

Finally, substituting for b_{max} and $P(b)$ into eqn 1.4 yields the (translational) energy-dependent reaction cross-section

$$\sigma_r(\epsilon_t) = \pi R_0^2\left(1 - \frac{\epsilon_0}{\epsilon_t}\right) \qquad \epsilon_t \geq \epsilon_0. \tag{4.3}$$

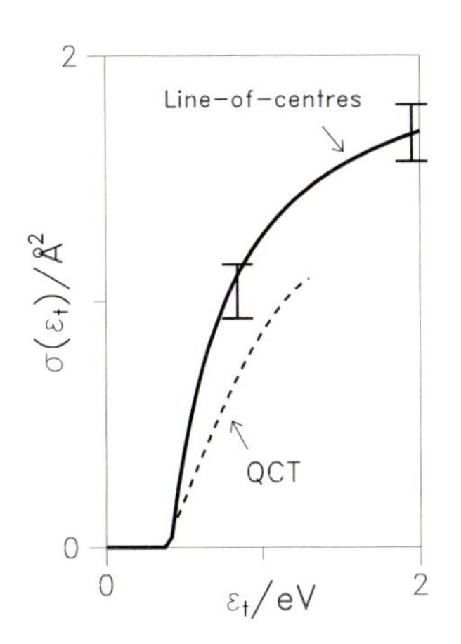

Fig. 4.1 Comparison of the experimentally and QCT derived $H + D_2$ excitation function with that obtained from the line-of-centres model.

QCT calculated and experimental cross-sections, $\sigma_r(\epsilon_t)$, for the $H + D_2$ reaction are shown in Fig. 4.1, where they are compared with the functional form predicted by the line-of-centres model. The experimental data were obtained using the pulsed laser pump–probe method. The model excitation function has been scaled (by varying R_0) to agree most closely with the experimental results. Although qualitatively correct, eqn 4.3 predicts a more rapid increase in cross-section with increasing collision energy than observed in the parameter-free QCT calculations.

The inaccuracies of the simple line-of-centres model are emphasized further if eqn 4.3 is used to evaluate the thermal rate constant, $k(T)$, by averaging over a Maxwell–Boltzmann distribution of collision energies using eqn 1.7. The result is

$$k(T) = \left(\frac{8k_BT}{\pi\mu}\right)^{1/2} \pi R_0^2\, e^{-\epsilon_0/k_BT},$$

which is identical to the simple collision theory expression for $k(T)$, eqn 1.3, provided the separation of the particles at the reaction barrier is equated with the sum of the reactant hard sphere radii, d. We have already seen that this equation yields A-factors in marked disagreement with experiment.

Significant improvements to the line-of-centres model can be made by allowing approximately for a steric effect.[32] In the modified line-of-centres model this is achieved by including an orientation dependent barrier, $\epsilon_0(\gamma)$.

Reactions without a barrier. For reactions which proceed on *attractive* potential energy surfaces without a barrier, the presence of a centrifugal barrier on the effective potential may still prevent reaction. Assuming, as in the line-of-centres model, that the reactant particles are structureless, eqn 4.1 may be used to evaluate the maximum impact parameter for reaction, provided the shape of the long-range attractive potential is known. The latter is often parameterized in the form (cf. eqn 2.3)

The procedure adopted below, which assumes that the reaction cross-section depends solely on the position and height of the centrifugal barrier on the long-range part of the effective potential energy surface, leads to what is commonly referred to as a *capture* cross-section, or rate constant.

$$V(R) = -\frac{C_n}{R^n},$$

where C_n is a constant, and $n = 4$ for ion–molecule reactions and $n = 6$ for reactions involving neutral species. The first task is to evaluate the position of the centrifugal barrier on the *effective* potential, $V_{eff}(R)$, which is the sum of the potential energy and the centrifugal energy introduced in Section 3.1:

$$V_{eff}(R) = \epsilon_t\frac{b^2}{R^2} - \frac{C_n}{R^n}.$$

$V_{\text{eff}}(R)$ has a maximum at $R = R_0$, where

$$R_0 = \left(\frac{n\,C_n}{2\,\epsilon_t\,b^2}\right)^{1/(n-2)}.$$

Applying eqn 3.4 with this value of R_0 provides an expression for the maximum impact parameter for which $\frac{1}{2}\mu\dot{R}^2 \geq 0$. As in the line-of-centres model, the opacity function for reaction is assumed to be unity for $b \leq b_{\text{max}}$, and the resulting expression for the excitation function is

$$\sigma_r(\epsilon_t) = \pi b_{\text{max}}^2 = \pi\left(\frac{n}{2}\right)\left(\frac{2}{n-2}\right)^{(n-2)/n}\left(\frac{C_n}{\epsilon_t}\right)^{2/n}. \qquad (4.4)$$

Thus, cross-sections for reactions proceeding over attractive potential energy surfaces are expected to decrease with increasing collision energy. The qualitative reason for this decrease is that, for a given impact parameter, higher collision energies lead to higher centrifugal barriers. Thus, as the collision energy increases, smaller values of b_{max} are necessary to ensure that the radial kinetic energy is sufficient to surmount the centrifugal barrier, as is evident from eqn 4.4.

Decreasing excitation functions with increasing collision energy are indeed found for ion–molecule reactions, and there is often reasonable agreement with the above *capture* model. Reactions involving neutral species which do not possess reaction barriers and proceed over attractive potential energy surfaces, such as $K + I_2$ and $O(^1D) + H_2$, also often display decreasing reaction cross-sections with increasing collision energy, although the precise functionality is sensitive to the shape of the potential energy surface at long range.

For reactions such as $K + I_2$, which, as we have seen, proceed *via* the harpoon mechanism, estimates of the reaction cross-section (and, in principle, of its collision energy dependence) can be made from a knowledge of the intersection region between the covalent and ionic surfaces (see Fig. 4.2). Suppose that the long range dependence of the covalent surface is dominated by the attractive van der Waals term (i.e. $V(R) = -C_6/R^6$). Then the crossing point, R_c, between the covalent and ionic curves is obtained when their potential energies are equal;

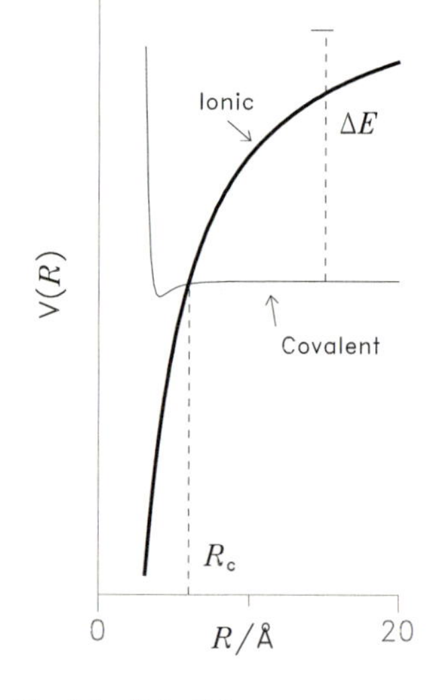

Fig. 4.2 Potential energy curves for a harpoon reaction.

$$-\frac{e^2}{4\pi\varepsilon_0 R_c} + \Delta E = -\frac{C_6}{R_c^6},$$

where the first term is the attractive Coulomb potential between two singly, oppositely charged species (e.g. $K^+ + I_2^-$) and $1/(4\pi\varepsilon_0) \sim 9 \times 10^9$ J C^{-2} m. ΔE is the asymptotic separation of the covalent and ionic potentials and is given by $\Delta E = E_{\text{IP}} - E_{\text{ea}}$, the difference between the ionization potential of the atom (A = K, in this example) and the electron affinity of the molecule ($BC = I_2$). Because the long range attractive potential on the covalent surface is small compared with the ionic potential, the right side of the above expression can be set equal to zero, and the separation of the reactants at the crossing point thus becomes

$$R_c \sim \frac{e^2}{4\pi\varepsilon_0 \Delta E}.$$

The crossing point provides an estimate of the maximum impact parameter for reaction. If we assume unit reaction probability for $b \leq b_{\max}$ (which is reasonable given the strong attraction between the reactants once they cross to the ionic potential energy surface), then the reaction cross-section will be

$$\sigma_{\rm r} = \pi R_{\rm c}^2.$$

For the alkali metal atoms, the ionization potentials decrease down the group (Li,..., Cs), and, therefore, the crossing radii and cross-sections for reactions with a given halogen molecule are expected to increase down the group. Furthermore, because the electron affinity is positive if electron attachment to the molecule is exothermic, molecules with large positive electron affinities possess large reaction cross-sections and a propensity for forward peaking differential cross-sections. By contrast, the electron affinity of CH_3I is slightly negative. The cross-section for the K + CH_3I reaction is, therefore, very small, and the reaction products are backward scattered in the CM frame (see Section 3.3).

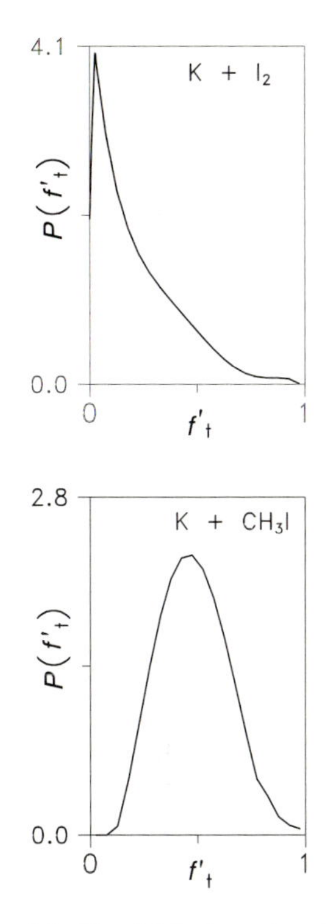

Fig. 4.3 The distributions of product kinetic energy releases in the K + I_2 and K + CH_3I reactions. f'_t is the fraction of the available energy released into translation.

Motion over the potential energy surface

The simple one-dimensional models presented above do not provide any insight into the relative importance of kinetic energy versus internal (vibrational or rotational) energy in promoting (or inhibiting) chemical reaction. Nor do they provide enlightenment about the origin of mode-specific disposal of excess energy in chemical reactions. An understanding of the motion of the nuclei over the full potential energy surface is required to rationalize such behaviour.

Energy disposal. The contrasting energy disposal data for the reactions K + I_2 and K + CH_3I was noted in Section 3.3. The data are replotted in Fig. 4.3, which shows the distribution of product translational energies in the two reactions. Because total energy is conserved (and is well defined in the crossed beam experiments), high product translational excitation correlates with low internal excitation of the reaction products, and vice versa. KI from the reaction K + I_2 is thus endowed with high internal excitation, while KI from K + CH_3I is generated with low internal excitation. The data may be accounted for qualitatively by *Polanyi's rules*,[33] which state that internal excitation in the products is enhanced by an increasingly early release of the reaction exoergicity in the entrance valley of the potential energy surface. The schematic trajectories drawn on the two collinear surfaces, shown in Fig. 4.4, illustrate this propensity. Because the surfaces are drawn for collinear geometries only, the internal excitation corresponds to product vibrational excitation. As the exoergic energy release occurs earlier in the entrance valley, trajectories are accelerated increasingly along the reactant valley, parallel to the x axis in Fig. 4.4. This coordinate corresponds to the bond length of the newly forming AB product, and as the early release of exoergicity increases, trajectories are unable to follow the potential contours and overshoot the minimum in the product valley of the potential energy surface, leading to preferential release of exoergicity into AB vibrational excitation. The harpoon reaction, K+I_2, with its large crossing radius and early energy release, conforms to this situation. For K + CH_3I reaction, by contrast, the energy is released when the reactants have approached to close proximity, and a larger

In the mass-scaled coordinates, these two reactions have skewing angles of $\sim 70°$ and $\sim 80°$: see Section 4.3.

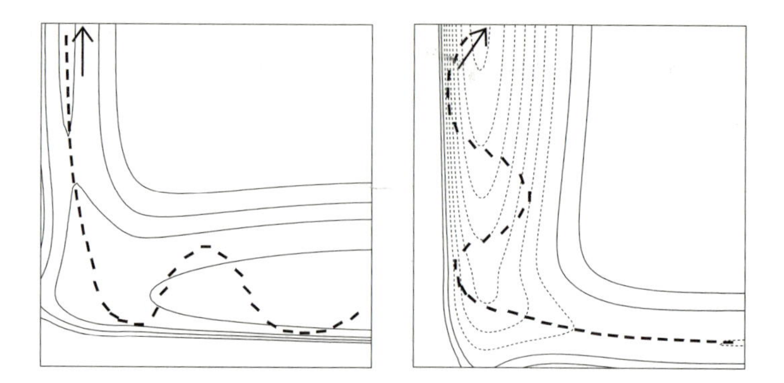

Fig. 4.4 Illustration of Polanyi's rules governing the disposal of excess energy on surfaces with late (left) and early (right) energy releases. In the *reverse* directions the barriers occur early (left) and late (right).

fraction of the reaction exoergicity is released as product translational excitation. Further examples are discussed in Section 4.3.

Energy utilization. If the trajectories shown in Fig 4.4 were run backwards (i.e. starting with AB + C), the path mapped out would be the same as those shown, but would provide information about the relative efficiencies of initial kinetic versus vibrational excitation in promoting reaction in the endothermic direction. Figure 4.4 suggests that reactions which possess *early* barriers (barriers in the reactant valley) will be promoted preferentially by initial translational excitation, while reactions with *late* barriers (barriers in the product valley) will be promoted preferentially by reactant vibrational excitation.

The reaction between K and HCl(v) provides an example of reaction over a surface with a late barrier, and in this case the relative efficiencies of vibrational and translational reagent excitation have been determined.[34] Vibrational excitation of HCl to $v = 1$ enhances the reaction cross-section by two orders of magnitude, while the same amount of energy in reactant translation produces only a modest enhancement.

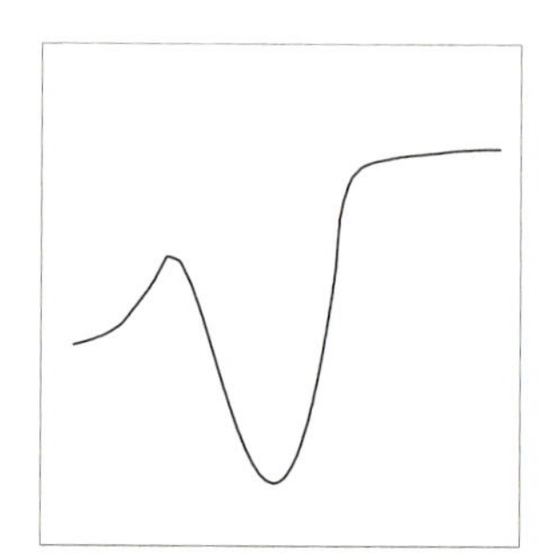

Fig. 4.5 A reaction profile which might lead to the statistical disposal of energy.

At fixed energy, the term 'statistical' means that every product quantum state has an equal probability of being populated.

Statistical disposal of energy

Up to this point emphasis has been placed on direct bimolecular reactions. These reactions display a dynamical bias, either in the way energy is used to surmount a barrier, or in the pattern of energy disposal. By contrast, for reactions which proceed via formation of an intermediate over potential energy surfaces with reaction profiles similar to that shown in Fig. 4.5, the energy disposal is often characterized by whether or not it is *statistical*. Such a reaction profile ensures relatively facile complex formation, but relatively slow decomposition to products. The absence of an exit barrier further ensures that energy is not released impulsively as the products separate. If the complex becomes trapped in the potential energy well for a sufficient length of time, its vibrational energy can become randomized amongst the different vibrational modes, via a process known as *intramolecular vibrational redistribution* (IVR) (see Section 5.3). If the fragments then separate without dynamical, mode-specific energy release, the internal state distributions in the reaction products may also appear randomized or statistical.

The statistical population may be calculated by evaluating the ratio of the number of product states associated with a specific product rotation-vibration level at a given total available energy, ϵ', divided by the total number of accessible states at that energy. If the state count is performed merely under the constraint of energy conservation, the statistical distribution is known as the *prior* distribution, $P_0(v', j')$. *Phase space* theory provides an alternative statistical distribution which, in addition to energy, also conserves total angular momentum. Such statistical distributions may be interpreted as the distributions expected in the absence of dynamical bias imposed by the forces described by the potential energy surface.

The prior distribution is comparatively easy to evaluate once it is recognized that product translational 'states' must be included in the state count. The density of translational states, the number of translational states per unit volume per unit energy, is given by

$$\rho_t(\epsilon_t) = \frac{\mu^{3/2}}{2^{1/2}\pi^2\hbar^3}\,\epsilon_t^{1/2}. \tag{4.5}$$

For the AB + C products of a triatomic reaction, the *prior* population in a given AB v', j' state is, therefore,

$$P_0(v', j') \propto (2j' + 1)\left[\epsilon' - \epsilon'_v - \epsilon'_r\right]^{1/2}.$$

The first term in this expression accounts for the degeneracy of state j', while the second term accounts for the translational density of states for the specific AB(v', j') products, born with translational energy $(\epsilon' - \epsilon'_v - \epsilon'_r)$, where ϵ'_v and ϵ'_r are the AB(v', j') vibrational and rotational energies.

The prior distribution is also used in *surprisal analysis* of product state distributions, which provides a measure of whether or not a distribution is statistical, as well as a convenient means of parameterizing data. Taking the example of vibrational populations, this analysis involves plotting the vibrational surprisal, defined as $\ln(P(v')/P_0(v'))$ (where $P(v')$ and $P_0(v')$ are, respectively, the measured and the prior vibrational population distributions), versus the fraction of the available energy, ϵ', released as vibrational excitation $f'_v = \epsilon'_v/\epsilon'$. Such plots are often linear, although rarely with the zero gradient predicted if the measured and prior distributions were identical. Interesting examples are provided by the surprisal analyses for the two reactions[35]

$$\mathrm{O + CS(Se) \longrightarrow CO}(v') + \mathrm{S(Se)},$$

which produce highly inverted, non-statistical vibrational state distributions. In spite of the rather different energetics for the two reactions, the surprisal plots in the two cases are almost the same, and both are approximately linear. The analyses suggest that the dynamics which lead to the vibrational population inversions in these two reactions are rather similar.

The surprisal analyses of these data are discussed at greater length by Pilling and Smith in *Modern gas kinetics*.

4.3 Kinematic constraints

A kinematic constraint is a constraint imposed purely on account of the reactant and product mass combination and the operation of the laws of conservation of (linear and angular) momentum and total energy. It is important to establish which observable properties of chemical reactions are determined by such mass factors, and which are under dynamical control.

Skew angle

When imagining classical trajectories on a two-dimensional slice of the potential energy surface (as exemplified by our discussion of Polanyi's rules), we envisaged the motion of the particles in terms of that of a point mass over a potential energy surface. However, this picture is only valid if the potential energy surface is drawn using the so-called *mass-scaled coordinates.*

A rereading of Section 3.2 of the preceding chapter might be helpful at this point.

The classical Hamiltonian for the three-atom A + BC reaction, eqn 3.9, may be rewritten explicitly in terms of (CM) velocity vectors, $\dot{\mathbf{R}}$ and $\dot{\mathbf{r}}$,

$$H = \tfrac{1}{2}\mu\left[\dot{\mathbf{R}}^2 + \left(\alpha^{1/2}\dot{\mathbf{r}}\right)^2\right] + V(R, r, \gamma),$$

where $\alpha = \mu_{\mathrm{BC}}/\mu$. For the *collinear* A + BC reaction this equation reads

$$H = \tfrac{1}{2}\mu\left[\dot{R}^2 + \left(\alpha^{1/2}\dot{r}\right)^2\right] + V(R, r, 0).$$

In this representation, the total kinetic energy of the system is written as a sum of two independent terms, with a common mass factor μ. The motion can be represented graphically by plotting the potential energy surface in the independent coordinates $x = R$ and $y = \alpha^{1/2}r$. Then the motion (trajectory) of the point of mass, μ, along the x coordinate would represent reactant approach, and motion along the y coordinate would represent vibrational motion of the diatomic reactant.

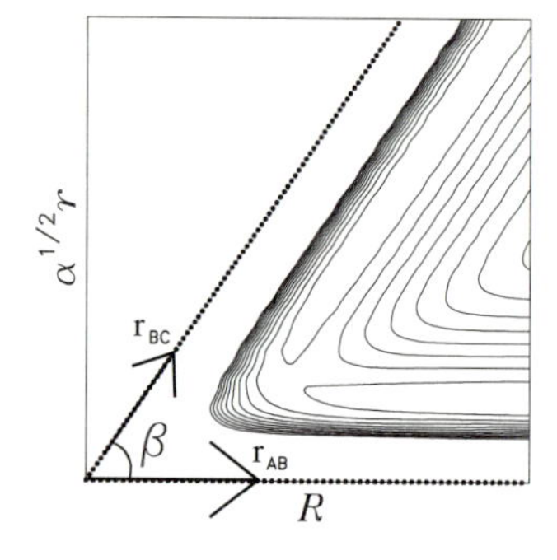

Fig. 4.6 The relationship between the mass-scaled and bond-axis coordinates.

Although the above form of the Hamiltonian provides the correct picture of the nuclear motion, our intuitive view of chemical reactions is in terms of bond lengths r_{AB}, r_{BC}, etc., rather than the coordinates R and $\alpha^{1/2}r$. For collinear reactant approach, the latter may be expressed in terms of the bond lengths as

$$x \equiv R = r_{\mathrm{AB}} + \frac{m_{\mathrm{C}}}{m_{\mathrm{BC}}} r_{\mathrm{BC}}$$

and

$$y \equiv \alpha^{1/2}r = \alpha^{1/2}\, r_{\mathrm{BC}}.$$

The mass scaled (x, y) coordinates of the bond lengths are given by $(r_{\mathrm{AB}}, 0)$ and $(\frac{m_{\mathrm{C}}}{m_{\mathrm{BC}}} r_{\mathrm{BC}}, \alpha^{1/2} r_{\mathrm{BC}})$.

Figure 4.6 illustrates schematically a potential energy surface plotted in the mass-scaled coordinates. Also shown are the bond length axes r_{AB} and r_{BC}. Although r_{AB} is found to lie along the x axis, r_{BC} is inclined at an angle β to the x axis, as illustrated, where

$$\cos^2\beta = \left[\frac{m_{\mathrm{C}}^2}{m_{\mathrm{C}}^2 + m_{\mathrm{BC}}^2\alpha}\right] = \frac{m_{\mathrm{A}}m_{\mathrm{C}}}{m_{\mathrm{AB}}m_{\mathrm{BC}}}.$$

Thus, if we are to view (collinear) reactions in terms of point masses moving over potential energy surfaces, we must scale and skew the bond axes through a *skew angle*, β. Furthermore, $\cos\beta$ is given simply in terms of a mass factor. The skewing angle will be close to 90° when the attacking (or departing) atom is much lighter than the diatomic ($\cos^2\beta \to 0$), but may be very small when the transferred atom, B, is much lighter than A or C ($\cos^2\beta \to 1$). Polanyi's rules, as described in Section 4.2, only apply in the former case, that of a light attacking atom. For reactions with skewing angles $\beta \ll 90°$, the pattern of energy disposal in exothermic reactions can be very different from that expected on the basis of Polanyi's rules, and the energy disposal data are most readily appreciated in the mass-scaled coordinate system. Generally, for exothermic reactions, as the skewing angle is reduced more of the energy is released as vibrational excitation of the diatomic product, as illustrated in Fig. 4.7.

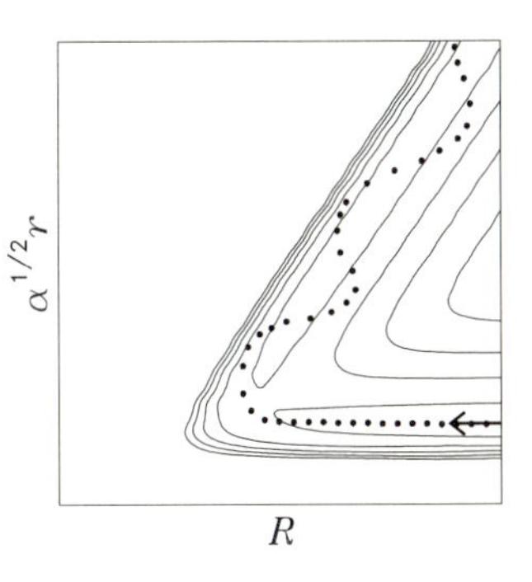

Fig. 4.7 Schematic trajectories on a surface with a small skew angle, β.

Angular momentum conservation

In order to describe chemical reactions fully, proper account of the reactant and product angular momenta must be taken. We have seen from Section 3.2 that there are two sources of angular momentum in three-atom reactions (disregarding electronic and nuclear spin angular momenta): the orbital angular momentum of the reactant pair, **L**, and the rotational motion of the reactant diatomic molecule, **j**. Similarly, for the products, we may identify *exit* orbital angular momentum, **L**′, and AB product rotational angular momentum, **j**′. In the absence of external forces, the *total* angular momentum of the system, **J**, must be conserved, and we may write

$$\mathbf{L} + \mathbf{j} = \mathbf{J} = \mathbf{L}' + \mathbf{j}' \qquad (4.6)$$

This equation should be contrasted with the angular momentum conservation constraint for atom–atom scattering, described in Section 3.1. In that case, the total angular momentum of the 'reactant' and 'product' atoms *is* the orbital angular momentum, $\mathbf{J} = \mathbf{L}$, and the latter must therefore be conserved in elastic atom–atom scattering.

Consider the rather special case of the F + H_2 reaction discussed in Section 3.3. This reaction is characterized, at moderate collision energy, by a rather small reaction cross-section, indicative of small impact parameter reactive collisions. This fact, together with the low reactant reduced mass, ensures that the magnitude of the reactant orbital angular momentum, $L = \mu v_r b$, is atypically small (for example, $L \leq 10\hbar$ for a collision energy of ~ 0.1 eV). The same is also true for the magnitude of the product orbital angular momentum, $L' = \mu' v_r' b'$, again partly reflecting the fact that $\mu' \sim \mu/2$ (at the same collision energy, L' is typically $\leq 10\,\hbar$). Low product orbital angular momentum is observed, particularly, for products born in higher vibrational states, because to conserve total energy these must be born with low product relative speeds, v_r'. The rotational angular momentum of the H_2 reactant is also very small. Even at room temperature, only a few rotational states of H_2 are thermally accessible. Thus, angular momentum conservation requires that $0 \leq j' \leq L' + J$. Typically only ten or so product rotational states are significantly populated at a collision energy of ~ 0.1 eV. Because only a few product rotational states are populated, the kinetic energy releases associated with HF born in different vibrational states are quite distinct, and, as a consequence, product vibrationally state resolved differential cross-sections can be observed in molecular beam experiments. This would not be possible if higher product rotational states were populated, because the kinetic energy releases associated with different product vibrational and rotational states would then overlap.

See Appendix A.2 for the expressions relating the classical to the quantal orbital angular momentum.

The vibrational distributions in the HF products of the F + H_2 reaction obtained from the molecular beam data are shown in Fig. 4.8 (similar data were also obtained in earlier IR chemiluminescence experiments[36]). Significant vibrational population inversion is observed, and on average, $\sim 65\%$ of the excess energy appears as product vibrational excitation and only $\sim 8\%$ as product rotation (the remainder necessarily resides in product kinetic energy). The selective disposal of energy is in qualitative accord with that expected for an exothermic reaction possessing an early barrier and a skew angle of only 47°, although quantitative agreement between calculated population distributions and experiment has only recently been obtained.[6]

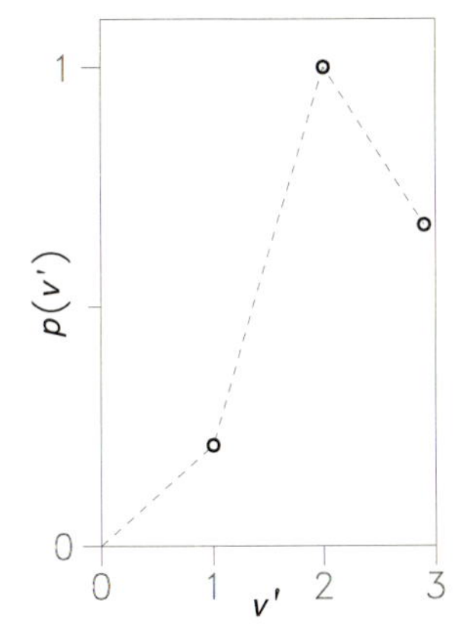

Fig. 4.8 HF product vibrational populations for the F + H_2 reaction.

Light-atom transfer reactions

Consider in more detail the class of reaction known as light-atom transfer reactions, which we may represent as

$$\mathrm{H} + \mathrm{LH} \longrightarrow \mathrm{HL} + \mathrm{H},$$

where H and L refer to heavy and light atoms. Such reactions have $\cos^2 \beta \to 1$ and, therefore, very small skew angles, which favour the release of excess energy (in the exothermic direction of the reaction) into vibrational excitation of the products (they also show a tendency to conserve kinetic energy). An example is the reaction

$$\mathrm{Cl} + \mathrm{HI} \longrightarrow \mathrm{HCl} + \mathrm{I},$$

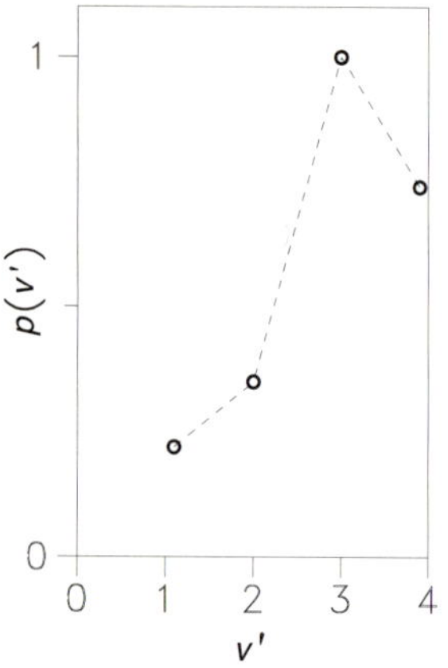

which has a skew angle of 10.8°. The experimentally derived vibrational distribution is shown in Fig. 4.9: the fractions of energy released into translation, rotation, and vibration are $f_t' \sim 0.15$, $f_r' \sim 0.15$ $f_v' \sim 0.70$. These features could be reproduced by classical trajectory calculations employing a *repulsive* potential energy surface (i.e. one with a late release of exoergicity), which, for a skew angle of 90°, would be expected to yield enhanced product translational energy release.[37]

The small reduced masses of the reactant and product LH diatomics means that their rotational energy level spacings are large. Thus, few reactant rotational states are populated at room temperature, and only comparatively few product states will be energetically accessible. For such reactions, therefore, one typically finds $j \ll L$ and $j' \ll L'$. Since J is conserved throughout reaction, and $L \sim J \sim L'$, there is a *propensity* for orbital angular momentum to be conserved during reaction,

$$\mathbf{L} \to \mathbf{L}'.$$

Fig. 4.9 The product HCl vibrational populations for the Cl + HI reaction.

This is particularly evident for light-atom transfer reactions proceeding over attractive potential energy surfaces, because they are characterized by large reaction cross-sections and, hence, large initial orbital angular momenta. Recall that if orbital angular momentum is conserved during reaction, and if the reaction proceeds via a long-lived complex, then the differential cross-section is expected to display symmetric peaks in the forward and backward directions (see Section 3.3).

Reactions involving light attacking or departing atoms

Now consider reactions at the other extreme of the kinematic spectrum, those involving attack of a light atom,

$$\mathrm{L} + \mathrm{HH} \longrightarrow \mathrm{LH} + \mathrm{H},$$

or a light departing atom,

$$\mathrm{H} + \mathrm{HL} \longrightarrow \mathrm{HH} + \mathrm{L}.$$

Both classes of reaction are characterized by $\cos^2 \beta \to 0$ and skew angles close to 90°. Because of the relative magnitudes of the reactant and product reduced masses in the two cases, the angular momentum disposal propensities may be written

$$\mathbf{j} \to \mathbf{L}'$$

and

$$\mathbf{L} \rightarrow \mathbf{j}',$$

respectively. We have encountered a reaction of the first class in Section 3.3, namely

$$D + I_2 \longrightarrow DI + I,$$

and the reactions of H/D atoms with the halogens (X_2) form a more general set of reactions conforming approximately to this kinematic limit. In these systems, a correlation is observed (most notably in QCT calculations on semi-empirical potential energy surfaces) between the percentage of the exothermicity released in the entrance valley, and the fraction of energy deposited in the internal rotational and vibrational modes of the product hydrogen halides, as anticipated on the basis of Polanyi's rules. Thus, in the $H + Cl_2$ reaction, energy is released when the H atom is in close proximity to Cl_2 and $f_t' \sim 0.5$, while for $D + I_2$ some 70% of the energy is released as product internal excitation, reflecting the earlier release of the exoergicity.[38]

The first propensity is only valid if the reactant orbital angular momentum is small compared with *j*. In spite of the low reactant reduced mass, μ, the magnitude of *L* is determined by $\mu v_r b$, and *L* is only likely to be small for reactions dominated by comparatively small impact parameter reactive collisions, at low collision energies.

The differential cross-section for the $D + I_2$ reaction was observed to display predominantly sideways scattering. One possible explanation for the behaviour may be a preference for sideways attack of the D atom to the I_2 bond (see Section 3.3). If this is the case, collisions will be more reactive if **j** is aligned parallel to the reactant relative velocity vector, since these collisions will be more likely to involve side-on attack than if **j** is aligned perpendicular to $\mathbf{v}_r$ (see Fig. 4.10). In this kinematic limit, **j** maps on to $\mathbf{L}'$, and, therefore, the latter too will be preferentially oriented parallel to the D atom beam direction. Finally, because the product relative velocity necessarily lies perpendicular to the product orbital angular momentum, a preference for sideways scattering of the DI products with respect to the D atom direction is predicted. By contrast, the $H + Cl_2$ and Br_2 reactions are observed to yield predominantly backward scattering, suggesting that these reactions proceed preferentially via collinear geometries.

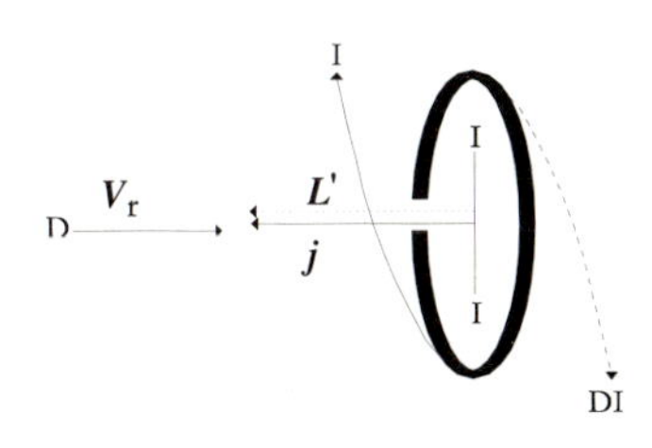

Fig. 4.10 The generation of sideways scattering in the $D + I_2$ reaction.

An example of a reaction involving a light departing atom is the exothermic harpoon reaction

$$Ba + HI \longrightarrow BaI(v', j') + H,$$

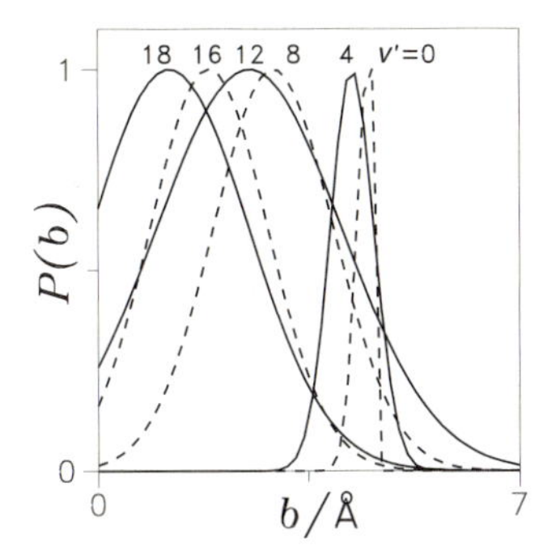

Fig. 4.11 The BaI(v') state resolved opacity functions, $P_{v'}(b)$, for the reaction Ba + HI.

which generates significant amounts of rotational and vibrational excitation in BaI. The importance of this kinematic class of reaction stems from the fact that the distribution of product rotational angular momentum is determined, via the angular momentum propensity given above, by the distribution of *reactive* reactant orbital angular momentum, *P*(*L*). At fixed collision energy, the latter is proportional to the opacity function, *P*(*b*). Thus, measurement of the rotational population distributions in specific vibrational levels of the reaction products allows direct determination of product vibrational state specific opacity functions. Such measurements have been made for the above reaction,[39] using the crossed molecular beam technique, coupled with laser-induced fluorescence probing of the rovibrational distribution in the nascent BaI products. The resulting opacity function data are plotted schematically in Fig. 4.11. Depending on the collision energy of the experiment, the maximum reactant orbital angular momentum, L_{max}, is determined either by the maximum value of j' allowed by energy conservation, or by the height of

This is the converse of the correlation identified above for the $D + I_2$ reaction, where the distribution of product orbital angular momentum is determined by the distribution of *reactive* reactant rotational angular momentum.

The importance of kinematic and angular momentum conservation constraints in chemical reactions is highlighted by the success of *kinematic models*.[40] The simplest of these is the *spectator stripping* model. Both this, and more sophisticated models, can be derived from the coordinate transformation equations given in Section 3.2 (see Levine and Bernstein's *Molecular reaction dynamics and chemical reactivity*).

the reactant centrifugal barrier. The BaI($v = 0$) products are generated in high product rotational states, suggesting an opacity function peaking very close to L_{max}. As the vibrational excitation in the BaI product is increased, the opacity functions peak towards smaller L or b values. Thus, highly vibrationally excited products arise from relatively low impact parameter collisions, while low vibrationally excited BaI products are generated from large impact parameter collisions. This correlation is primarily dictated by the kinematics and energetics of the reaction. The dynamical role played by the potential energy surface is in governing the *vibrational* populations of the BaI product.

4.4 Case studies

This and other examples are discussed in a review article by C.B. Moore and I.W.M. Smith.[41]

In recent years, very detailed experimental and theoretical studies have been undertaken of the reaction

$$D + H_2(v, j) \longrightarrow DH(v', j') + H,$$

as well as of the other isotopic variants of the H + H_2 system. In one such study[42], relative state-to-state reaction cross-sections (populations) were obtained by excitation of H_2 to specific rotational states of $v = 1$, using a non-linear laser technique known as stimulated Raman pumping. The hot D atom reactants were generated using photolysis of DI, and the rovibrational state population distributions in the HD products were determined by REMPI. Figure 4.12 compares the DH vibrational populations obtained for H_2 molecules in low rotational states of $v = 0$, and the $j = 1$ level of $H_2(v = 1)$. The data suggest that vibrational excitation in this system is *adiabatic*; i.e. vibrational excitation of the reactants yields enhanced vibrational excitation of the products. The population distributions compare favourably (although not perfectly at the highest of the collision energies studied) with the predictions of *ab initio* quantum mechanical calculations.

This excitation technique transfers about 30% of the $v = 0$ population to $H_2(v = 1)$. Because both vibrational states are populated, and both yield HD products, the reactant vibrational state specific energy disposal data are obtained by subtracting the signal obtained without $H_2(v = 1)$ from that obtained when both reactant vibrational levels are significantly populated.

Another reaction which has been investigated with full reactant and product energy level specificity is[43]

$$O(^3P) + HCl(v, j) \longrightarrow OH(v', j') + Cl(^2P).$$

The reaction is approximately thermoneutral, and possesses a sufficiently large barrier that the rate coefficient for reaction with HCl($v = 0$) is very small compared with that for higher vibrational states. HCl reactants were prepared in $v = 2, j = 1, 6, 9$ levels, using direct IR laser excitation, and the OH products were probed using LIF. The ground state oxygen atoms were generated by pulsed photolysis of NO_2. This light-atom transfer reaction possesses a small skew angle ($\beta \sim 17°$), and a significant fraction of the excess energy is released as OH vibrational excitation (the fractional energy releases into OH vibration and rotation are $f'_v \sim 0.4, f'_r \sim 0.3$). The vibrational distribution is inverted, and is consistent with the efficient transfer of reactant into product vibrational excitation. Such behaviour is expected for a light-atom transfer reaction, because the small skew angle ensures that the reactant vibrational coordinate is nearly parallel with that of the product in the mass-scaled coordinate system (see Fig. 4.13).

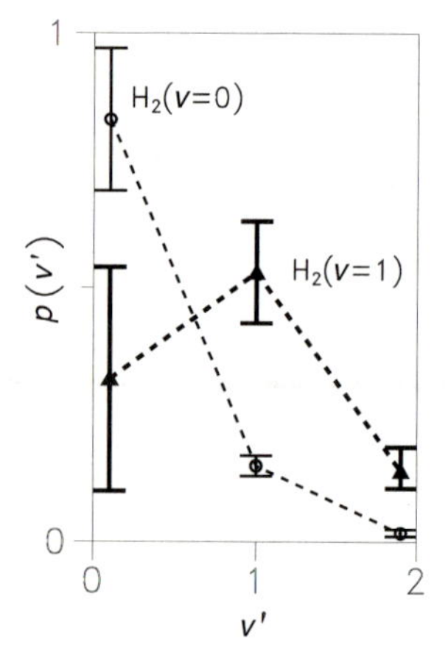

Fig. 4.12 The product HD vibrational populations subsequent to D atom reaction with H_2 in $v = 0$ and $v = 1$.

The particular emphasis in the study was the role played by *reactant* rotational excitation. Significantly, the fraction of energy released as product OH rotation was found to be insensitive to the reactant rotational state (although the OH rotational distributions do shift slightly to higher j' as the HCl rotational energy is increased). Since product rotational excitation does not originate from the rotational excitation of the reactants, an alternative source for the excitation was proposed, namely that the reaction proceeds preferentially via slightly bent (rather than linear) configurations in the barrier region. Therefore, product rotational excitation is generated by the repulsive release of energy at the barrier, which even for a slightly bent configuration is sufficient to produce the observed fragment rotational excitation.

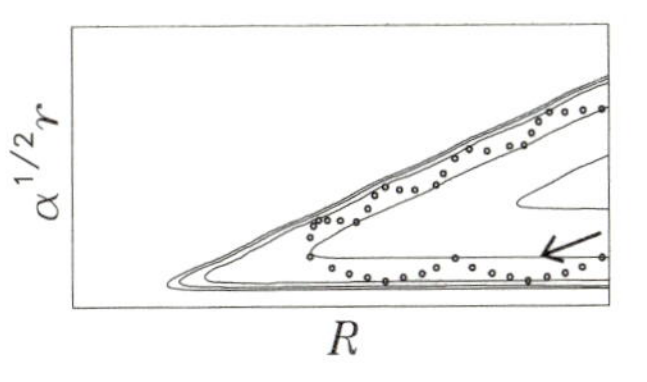

Fig. 4.13 Schematic trajectories for the O + HCl reaction, illustrating the efficient transfer of reactant to product vibrational excitation.

A rather dramatic example of vibrational mode specificity is provided by the reaction

$$H + HOD(v_1, v_2, v_3) \longrightarrow OH(OD)(v') + H_2(HD).$$

Relative rate coefficients for reactants in different vibrational states have been obtained by a number of researchers, and the reaction, together with other isotopic variants, has been the subject of review.[44] Selective excitation of H–OD or HO–D vibrational modes leads to preferential cleavage of the excited bond, yielding, with high efficiency, either OD or OH products. The reaction is endothermic in the forward direction, and the minimum energy path is thought to pass through planar HHOH geometries, with a (minimum) barrier occurring for near-collinear H–H–O configurations. Vibrational excitation in one of the HOD bonds is thought to lower and open the reaction barrier; i.e. to open the reactive cone of acceptance (see Section 3.4). Interestingly, the *reverse* reaction, $OH(v_1) + H_2(v_2)$, also displays pronounced vibrational mode specificity.[45] Excitation of the 'spectator' OH vibration to $v_1 = 1$ leads to little change in the reaction rate coefficient, compared to that employing ground vibrational state reactants. By contrast, vibrational excitation to $v_2 = 1$ in the H_2 bond (i.e. the bond to be broken) leads to a factor of 100 enhancement in rate coefficient.

The data also support the notion that vibrational excitation in the reactant HOD molecule is not efficiently randomized on the timescale of the experiment via the process of IVR; see Section 5.3.

As a final example, we return briefly to the exothermic reaction

$$O(^1D) + H_2 \longrightarrow OH(v', j') + H,$$

which, at low collision energies, is believed to occur via an insertion mechanism leading to a short-lived H_2O intermediate (see Section 3.3). The internal state distribution in the OH products have been determined via LIF, subsequent to $O(^1D)$ atom generation by photolysis of either O_3 or N_2O.[46] A significant fraction of the excess energy ($f'_r \sim 0.3$) is deposited as OH rotational excitation, in addition to vibrational excitation ($f'_v \sim 0.4$), suggesting some memory in the reaction products of the highly bend-excited, transitory H_2O intermediates. One consequence of the mechanism for production of OH rotation (evident from QCT calculations) is that, for high rotational states in $OH(v' = 0)$, the rotational angular momentum vector lies antiparallel to the product orbital angular momentum vector, as depicted in Fig. 4.14. Such vector orientation properties are another aspect of stereochemistry, discussed in Section 3.4.

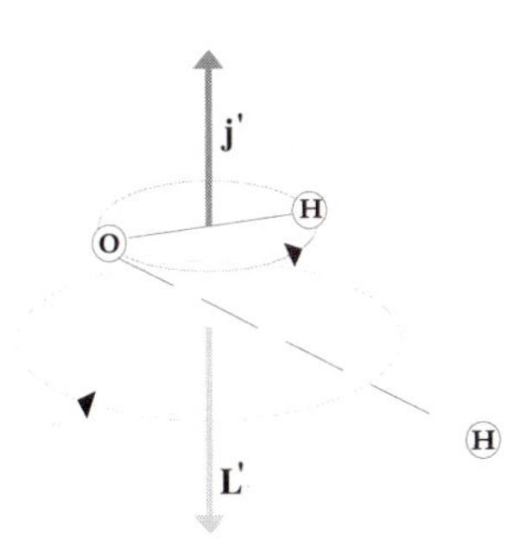

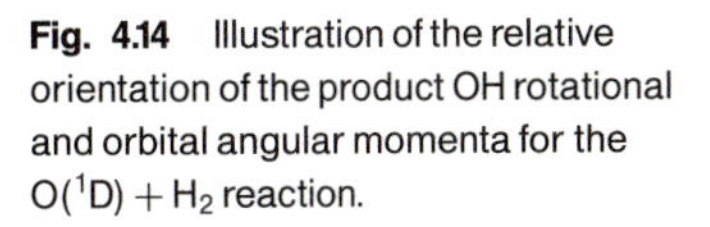

Fig. 4.14 Illustration of the relative orientation of the product OH rotational and orbital angular momenta for the $O(^1D) + H_2$ reaction.

Analogous behaviour is observed when water is photolysed *via* a linear excited electronic state. The sudden bent-to-linear geometry change induced by photon absorption leads to excited rotational OH products, with the product rotational angular momentum vector aligned antiparallel to that of the product orbital angular momentum.

5 Microcanonical rate coefficients

5.1 The cumulative reaction probability

Motion along the reaction coordinate

The variation of the potential energy along the minimum energy path is very different from that orthogonal to it (see Section 2.2). Motions along the latter may be thought of as conventional (symmetric) stretching and bending vibrations, whereas the motions along the former correspond to translational motion over the barrier. We focus, initially, on the one-dimensional (1-D) motion along the reaction coordinate.

The rate, $d\nu$, at which the reactants (A and BC) approach one another along the reaction coordinate (i.e. the reactant flux—the number of reactant pairs approaching per unit length per unit time), at relative velocities between v_r and $v_r + dv_r$, is given (in 1-D) by

$$d\,\nu = v_r f(v_r)\,[\mathrm{A}]\,[\mathrm{BC}]\,dv_r,$$

where $f(v_r)$ is the 1-D Maxwell–Boltzmann distribution of relative speeds. The reaction rate, $d\nu_r$, is the probability of reaction, $P(\epsilon_t)$, at a given collision energy, ϵ_t, times the rate at which the reactants approach:

$$d\,\nu_r = P(\epsilon_t)\,v_r f(v_r)\,[\mathrm{A}]\,[\mathrm{BC}]\,dv_r.$$

Hence the thermally averaged reaction rate coefficient may be written

$$k(T) = \int_0^\infty P(\epsilon_t)\,v_r f(v_r)\,dv_r,$$

which, in 1-D, has units of m molecule^{-1} s^{-1}. This may also be expressed purely in terms of the initial translational energy, $\epsilon_t = \frac{1}{2}\mu v_r^2$, as

$$k(T) = \int_0^\infty P(\epsilon_t) f(\epsilon_t)\,\frac{d\epsilon_t}{\mu},$$

where the 1-D Maxwell–Boltzmann distribution of collision energy is given by

$$f(\epsilon_t) = \frac{\mu}{hq_t}\,e^{-\epsilon_t/k_B T},$$

and q_t, the 1-D translational partition function *per unit length*, represents the number of thermally accessible translational states per unit length at temperature, T:

$$q_t = \left(\frac{2\pi\mu k_B T}{h^2}\right)^{1/2}.$$

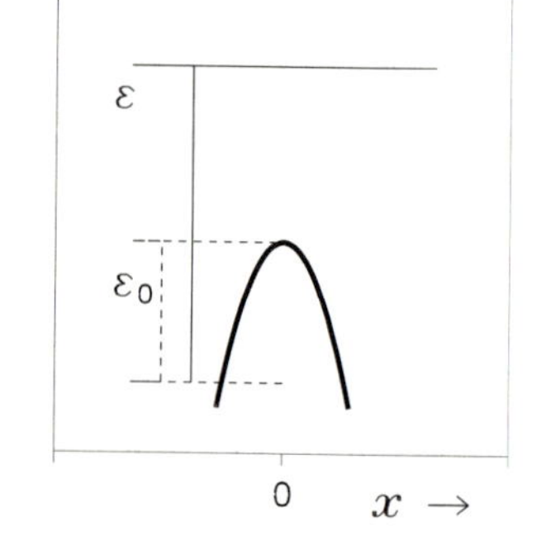

Fig. 5.1 1-D parabolic potential curve.

Substituting $f(\epsilon_t)$ into the above expression for $k(T)$ yields

$$k(T) = \frac{1}{h\,q_t}\int_0^\infty P(\epsilon_t)\,e^{-\epsilon_t/k_B T}\,d\epsilon_t. \tag{5.1}$$

On realistic potential energy profiles, the initial collision energy is equal to the total energy, and ϵ_t may be replaced by ϵ in the above equation.

Consider the 'reaction' probability for 1-D motion (in the x direction, say) over a barrier of height ϵ_0. Reaction corresponds to passage of a particle over (or through) the barrier. Assume the barrier to be an inverted parabola in x (i.e. $V(x) = \epsilon_0 - kx^2/2$ as shown in Fig. 5.1). Classically, the reaction probability (the probability that a particle, initially at $x < 0$ moving in the $+x$ direction, crosses the barrier) will be unity provided the total energy, ϵ (the sum of the kinetic and potential energies), exceeds the barrier height, ϵ_0. In terms of the total energy, ϵ, $P(\epsilon)$ is a step function, as illustrated in Fig. 5.2, and can be written as

A somewhat more realistic model potential energy curve for motion along the reaction coordinate is the Eckart barrier, discussed qualitatively by Atkins and Friedman in *Molecular quantum mechanics*.

$$P(\epsilon) = 1 \qquad \epsilon > \epsilon_0,$$

and

$$P(\epsilon) = 0 \qquad \epsilon \le \epsilon_0.$$

Quantum mechanically, a particle of mass μ approaching the barrier from the left hand side of Fig. 5.1 can be transmitted through the barrier or reflected back in the $-x$ direction. For the parabolic barrier problem, the reflection ($R(\epsilon)$) and transmission ($T(\epsilon)$) probabilities can be expressed analytically

$$R(\epsilon) = \frac{1}{1 + e^{2\pi\epsilon'}}$$

and

$$T(\epsilon) \equiv P(\epsilon) = \frac{1}{1 + e^{-2\pi\epsilon'}}, \tag{5.2}$$

where

$$\epsilon' = \frac{\epsilon - \epsilon_0}{h\nu^*},$$

and ν^*, the *imaginary vibrational frequency*, is defined as

$$\nu^* = \frac{1}{2\pi}\sqrt{\frac{k}{\mu}}.$$

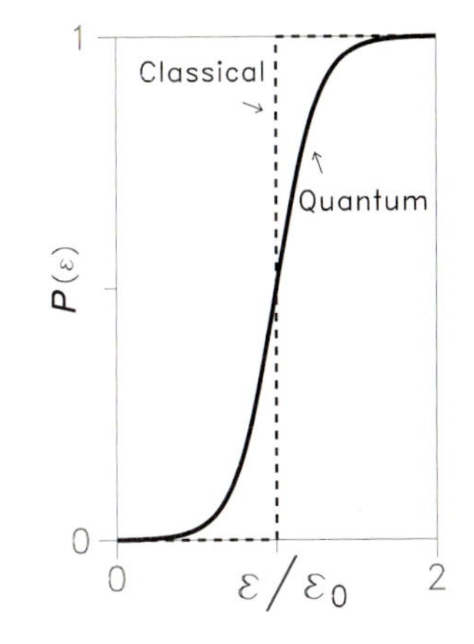

Fig. 5.2 Classical and quantum mechanical $P(\epsilon)$ for 1-D parabolic potential curve.

Thus, as shown in Fig. 5.2, $P(\epsilon)$ is a monotonically increasing function of ϵ and approaches the classical value of 0 when $\epsilon \ll \epsilon_0$, and of 1 when $\epsilon \gg \epsilon_0$. For energies below the barrier, the barrier is not totally opaque (i.e. the particle can *tunnel* through the barrier), and above the barrier, the barrier is not totally transparent (i.e. the particle can be reflected by the barrier even though the total energy exceeds the barrier height). Note that the quantum result tends to the classical one as μ increases, and k, the force constant, decreases. Thus quantum mechanical tunnelling will be most important for light particles passing through thin barriers.

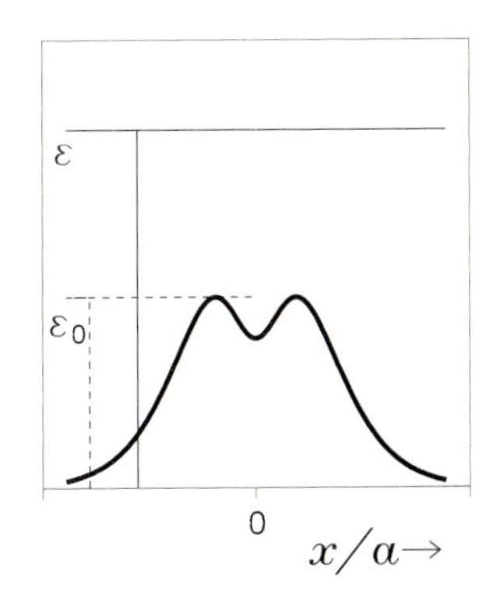

Fig. 5.3 1-D potential curve with a well.

Another illustrative example is provided by the one-dimensional motion of a particle over the barrier shown in Fig. 5.3, which has a small well in it, and

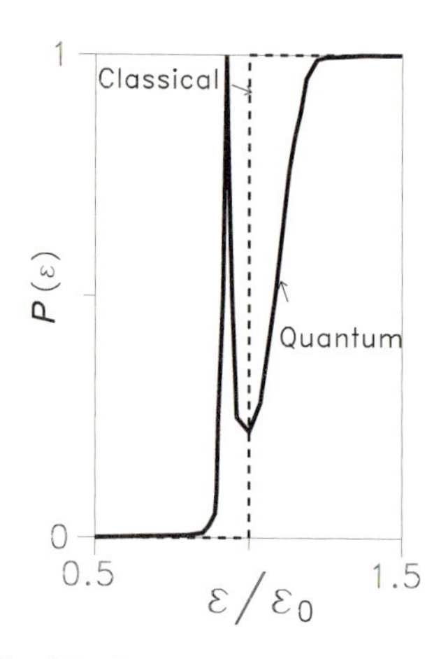

Fig. 5.4 Quantum mechanical $P(\epsilon)$ for 1-D potential with well.

has the form

$$V(x) = V_0\left[1/2 + (x/a)^2\right]\text{sech}^2(x/a).$$

The classical reaction probability for this potential energy profile is exactly the same as for the parabolic barrier. If the particle has enough (kinetic) energy to surmount the first barrier, it must have enough energy to surmount the second, and hence to 'react'.

By contrast, the quantum mechanical transmission probability, $P(\epsilon)$, shown in Fig. 5.4, is remarkably different from that obtained in the absence of a well.[47] Superimposed on the monotonically increasing transmission probability with increasing ϵ is a peak or *resonance*. Depending on the depth of the well and the particles' reduced mass, the potential well may support *quasi-bound* vibrational levels. When the total energy is close to that of a quasi-bound level, the reaction probability is observed to increase abruptly. For this particular (symmetrical) potential function, $P(\epsilon)$ increases to unity; the barrier becomes completely transparent in the region of the resonance. The phenomenon is known as *resonant tunnelling*. The behaviour is analogous to that arising from tunnelling through centrifugal barriers in effective potentials: see Section 3.1.

Motion in multiple dimensions

Now consider an A + BC reaction in its full dimensionality, and assume that the reactants are in thermal equilibrium. They approach one another in specific quantum states, and eqn 5.1 applies to each reactant quantum state, n, weighted by the Boltzmann population for the molecules in that state:

These reactant states, in addition to including the reactant BC internal states, also include the reactant orbital angular momentum quantum states, which are labelled by a quantum number ℓ and have a degeneracy $2\ell + 1$ (see Appendices A.2 and A.3).

$$k(T) = \frac{1}{h\,q_t}\int_0^\infty \sum_n P_n(\epsilon)\,\frac{g_n e^{-\epsilon_n/k_B T}}{q_{A:int}\,q_{BC:int}}\,e^{-\epsilon_t/k_B T}\,d\epsilon_t, \tag{5.3}$$

where g_n is the degeneracy of state n, and the partition functions are given by

$$q_{A:int} = q_{A:el}$$

$$q_{BC:int} = q_{BC:rot}\,q_{BC:vib}\,q_{BC:el},$$

and (since we are now in 3-D)

$$q_t = \left(\frac{2\pi\mu k_B T}{h^2}\right)^{3/2}.$$

The internal partition function for atom A is simply the electronic partition function, which is usually given by the electronic degeneracy of the ground electronic state. The form of the internal partition function of BC assumes that the reactant degrees of freedom are separable.

Eqn 5.3 is usually written more compactly as

$$k(T) = \frac{1}{hq_r}\int_0^\infty N(\epsilon)\,e^{-\epsilon/k_B T}\,d\epsilon, \tag{5.4}$$

(which is the same as eqn 1.13) where ϵ is the total energy (the sum of the reactant kinetic and internal energies)

$$\epsilon = \epsilon_t + \epsilon_n,$$

q_r $(= q_t\,q_{A:int}\,q_{BC:int})$ is the total reactant partition function, and

$$N(\epsilon) = \sum_n g_n\,P_n(\epsilon).$$

Therefore, $N(\epsilon)$ is the sum of the reaction probabilities (lying between 0 and 1) for all reactant states. It can be interpreted, more loosely, as the number of *reactive* reactant states at energy, ϵ (see Section 1.4).

Quantum mechanically calculated cumulative reaction probabilities, $N(\epsilon)$, are illustrated for the reactions $H + H_2$[48] and $F + H_2$[6] in Fig. 5.5 and 5.6. The 'staircase' type structure in $N(\epsilon)$, which is common to both reactions, is reminiscent of the one-dimensional classical behaviour shown in Fig. 5.2. Note also the comparatively sharp resonance-type structure observed for the $F + H_2$ reaction at low energies: close to threshold, the $N(\epsilon)$ calculations suggest the presence of local minima (or wells) on the potential energy surface, which lead to peaks in $N(\epsilon)$ analogous to that shown in Fig. 5.4. Detailed analysis suggests that the majority of the sharp features in Fig. 5.6 can be associated with quasi-bound levels in the *product* valley, arising from a weak, long range H—HF van der Waals minimum in the potential energy surface. The peaks are associated specifically with van der Waals H—HF stretching vibrations, in which the HF molecule is in $v' = 3$.[20]

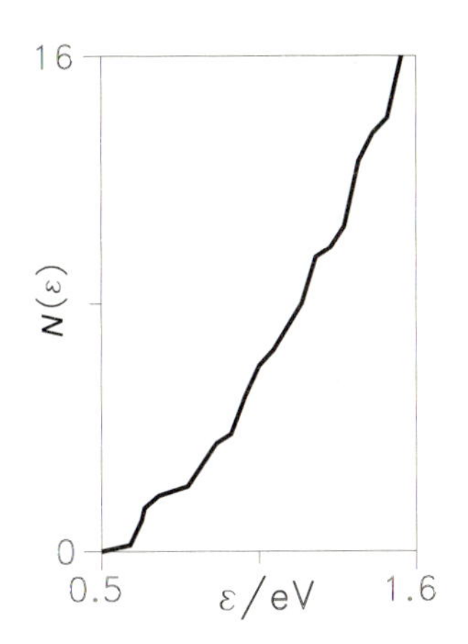

Fig. 5.5 Calculated cumulative reaction probability, $N(\epsilon)$, for the reaction $H + H_2$ for $J = 0$. The *total* angular momentum, J, is restricted to 0 in both these plots.

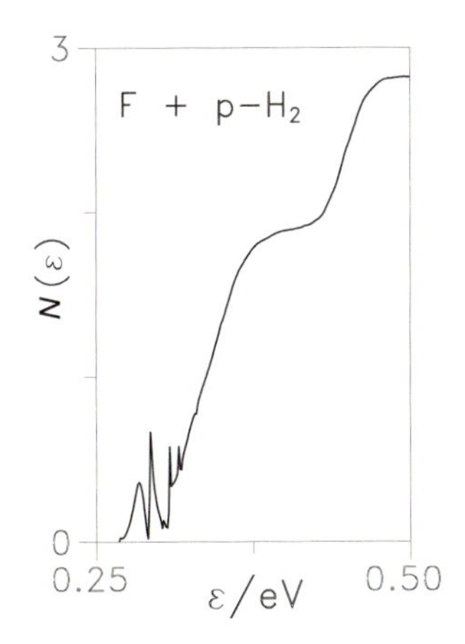

Fig. 5.6 Calculated cumulative reaction probability, $N(\epsilon)$, for the reaction $F + H_2$ for $J = 0$.

5.2 Transition state theory (TST) and $N^\ddagger(\epsilon)$

Eqn 5.4 is the exact expression for the *thermal* rate coefficient. However, the exact calculation of $N(\epsilon)$ by quantum mechanical means is computationally intensive, and is still only feasible for relatively light, three- or four-atom systems. It is desirable to explore alternative, approximate methods for calculating $N(\epsilon)$. In the following, we again treat the three-atom reaction

$$A + BC \longrightarrow ABC^\ddagger \longrightarrow AB + C.$$

The transition state (TS)

The transition state ($ABC^\ddagger$) may be thought of as a bottleneck somewhere on the potential energy surface, through which reactants must pass if they are to form products. More precisely, it is a *dividing surface* (for example, the slice in the collinearly constrained potential energy surface shown in Fig. 5.7), which separates reactants from products.

The assumptions of transition state theory are that

(1) at fixed temperature, the reactants are in thermal equilibrium, or, at fixed energy, all reactant states are equally accessible (see Section 5.3);
(2) motion along the reaction coordinate (the minimum energy path) is separable from motion orthogonal to it;
(3) motion along the reaction coordinate is classical;
(4) motion along the minimum energy path is direct.

In the version of transition state theory presented here, the (bound) motion orthogonal to the MEP is treated quantum mechanically.

Of these, the key dynamical assumption of transition state theory is that of 'direct dynamics' in assumption 4, which is often referred to as the *no-recrossing rule*. Trajectories which cross the transition state dividing surface from the reactant side are assumed to go on to form products and cannot be reflected back into the reactant valley.

The sum of transition states, $N^\ddagger(\epsilon)$

At the transition state (the critical geometry of no return), the second of the above assumptions allows one to write the total energy in the separable form

$$\epsilon = \epsilon_t^\ddagger + \epsilon_n^\ddagger + \epsilon_0,$$

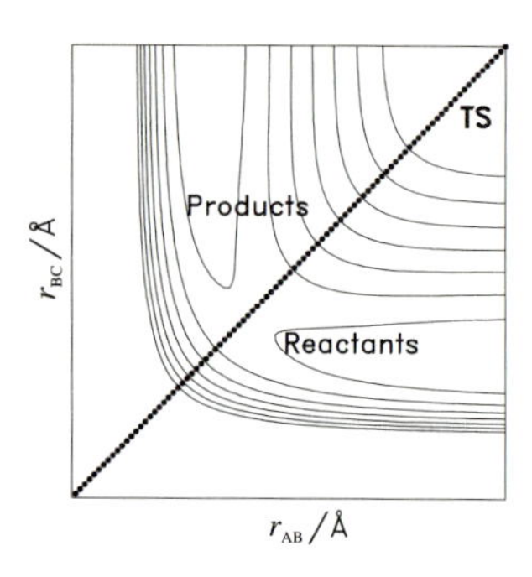

Fig. 5.7 The transition state dividing surface separating reactants from products.

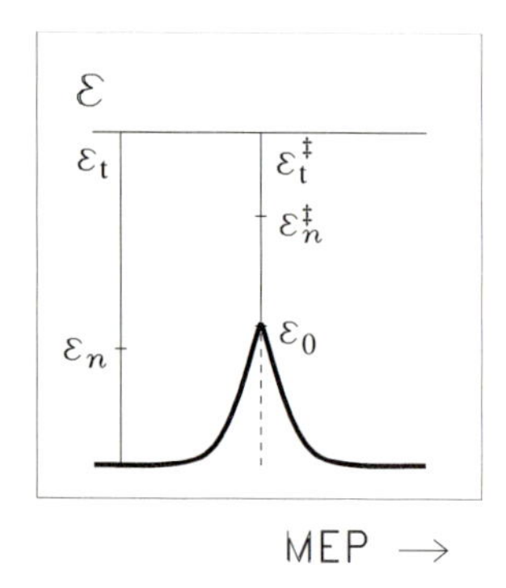

Fig. 5.8 The energy levels of the transition state.

$3N$–3 degrees of freedom are required to define the rotational and vibrational motion of the transition state. One degree of freedom corresponds to the reaction coordinate, and the remaining $3N$–4 correspond to the bound rovibrational coordinates of the transition state species.

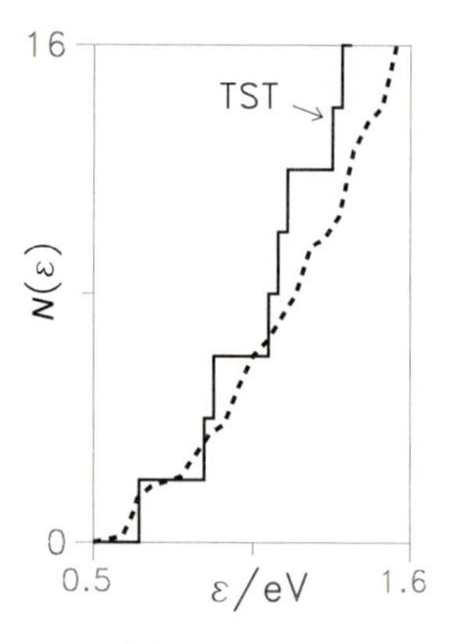

Fig. 5.9 $N^{\ddagger}(\epsilon)$ calculated assuming separable, harmonic vibrational motion for the $H + H_2$ reaction. The data are compared with the exact $N(\epsilon)$ shown in Fig. 5.5.

where ϵ_0 is the *difference in zero-point energies* between the reactant and transition state (see Fig. 5.8). If motion along the reaction coordinate is treated classically at the transition state, then each of its internal states constitutes a threshold to reaction, and the reaction probability for each of these states must be a step function,

$$P_n(\epsilon) = 1 \qquad \text{if } \epsilon > \epsilon_n^{\ddagger} + \epsilon_0,$$

and zero otherwise, where we have assumed that, once formed, the transition state species must go on to yield products. As the total energy increases and exceeds each threshold, the transition state estimate of the cumulative reaction probability will increase by unity (cf. the classical one-dimensional motion shown in Fig. 5.2); i.e. $N^{\ddagger}(\epsilon)$ can be expressed as

$$N^{\ddagger}(\epsilon) = \sum_n g_n^{\ddagger},$$

where the sum runs over all levels for which $\epsilon > \epsilon_n^{\ddagger} + \epsilon_0$. Within transition state theory, the cumulative reaction probability will thus equal the number of available transition states below the energy, ϵ. It will possess a 'staircase' structure with increasing total energy, where each step (of unit height) corresponds to the opening of a new threshold to reaction at the transition state. In transition state theory, the cumulative reaction probability (the sum of reactive reactant states) is approximated by the sum of transition states.

For reactions with large barriers, the TS is best placed at the top of the reaction barrier, since the barrier constitutes the smallest bottleneck through which reactants must pass to form products (see the following section). The energy levels of the transition state could be calculated by solving the Schrödinger equation for the bound motions in the $3N$–4 degrees of freedom of the transition state molecule. In many applications, however, it is assumed that the vibrational modes are harmonic and the rotational motion is that of a rigid rotor, as illustrated by the following calculation.

Consider the $H + H_2$ reaction once more. We will confine our attention to calculating $N^{\ddagger}(\epsilon)$ for zero total angular momentum, $J = 0$. To simplify the calculation, assume that the motion at the TS is harmonic with stretch and doubly degenerate bend frequencies of 2058 and 909 cm^{-1} (these frequencies are derived from the same PES employed in the exact calculation of $N(\epsilon)$). To evaluate $N^{\ddagger}(\epsilon)$, we simply count the number of states below a given energy ϵ, taking into account the potential energy barrier of 0.418 eV and the TS zero-point energy of 0.24 eV. Note that $N^{\ddagger}(\epsilon)$ must be multiplied by a factor of 2 to account for the fact that either of the two H atoms in the H_2 reactant can be abstracted, and that the total energy in this example is defined to include the reactant H_2 zero-point energy. (To perform the state counting in this example it is necessary to account for the *vibrational* angular momentum in excited bending levels. For total angular momentum $J = 0$, it turns out that only even quanta in the bend can contribute, and *only one* of the degenerate bending levels with even quanta of excitation has zero angular momentum.) The data are listed in Table 5.1 and shown in Fig. 5.9. The staircase structure in $N^{\ddagger}(\epsilon)$ roughly mirrors that observed in the exact $N(\epsilon)$ data shown in Fig. 5.5, although the steps in the latter appear smoothed by quantum mechanical tunnelling. With such a simple calculation, the location of the steps is not expected to be precise: the exact $N(\epsilon)$ values are somewhat smaller than the

Table 5.1 Sum of $H_3^{\ddagger}$ transition 'states' with zero total angular momentum for the first 10 states, calculated on the assumption of independent harmonic motion. n_s and n_b are the number of stretching and bending quanta.

ϵ/eV	(n_s, n_b)	$N^{\ddagger}(\epsilon)$
0.66	0,0	2
0.88	0,2	4
0.91	1,0	6
1.10	0,4	8
1.14	1,2	10
1.17	2,0	12
1.33	0,6	14
1.36	1,4	16
1.39	2,2	18
1.43	3,0	20

TS estimates, which is, in part, a consequence of employing the harmonic and separable mode assumptions.

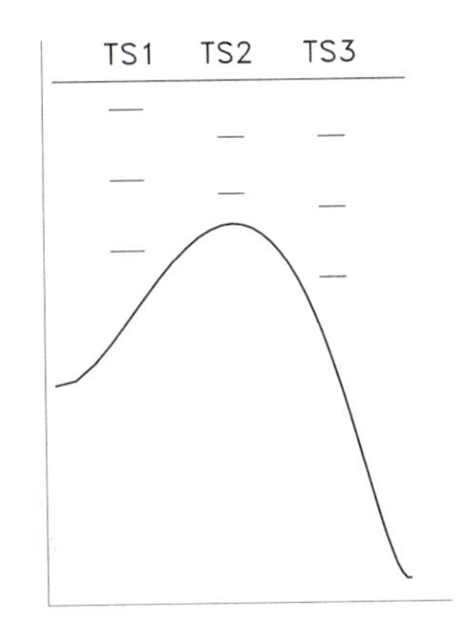

Fig. 5.10 States of the TS for three different locations of the dividing surface.

The 'no-recrossing' assumption

The TS dividing surface can be located anywhere on the potential energy surface, although some positions will be more appropriate than others. Figure 5.10 illustrates how the set of transition 'states' might vary with TS dividing surface location. Varying the transition state position leads to changes in the energy level spacings: different transition state locations correspond to different slices through the potential energy surface and, hence, to different vibrational force constants and moments of inertia.

A more general criterion for locating the best TS, particularly useful for reactions without a barrier, is to find the location which minimizes the number of transition 'states', $N^{\ddagger}(\epsilon)$. This choice provides the lowest estimate of $k(T)$. For reactions with large barriers the minimum in the number of transition states at a given energy coincides with the barrier location, because much of the energy at the barrier is fixed in the form of potential energy. Minimizing the number of transition states is desirable because it minimizes the effect of recrossing trajectories. The fraction of trajectories which recross the TS depends on the form of the potential energy surface and on the kinematic skewing angle introduced in Section 4.3. For surfaces with small skewing angles, trajectories are more prone to cross the TS in the barrier region, but are then unable to turn the sharp corner into the product valley and, instead, bounce off the repulsive wall of the potential and recross the TS back to reactants (see Fig. 5.11). Such recrossing trajectories lead to a lowering of the classical rate coefficient with respect to that obtained via TST (which effectively counts trajectories which recross as reactive). The best location of the TS is therefore the one which gives the lowest estimate of $N^{\ddagger}(\epsilon)$, and hence $k(T)$.

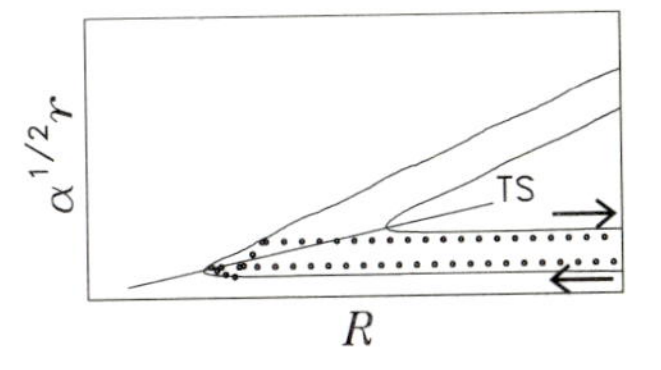

Fig. 5.11 Recrossing trajectories on a potential energy surface with a small skew angle.

Recrossing is usually a minor problem for reactions with high barriers, at low total energies, but can become more significant as the energy is raised.

In microcanonical *variational* transition state theory one searches for the best location of the transition state dividing surface by finding the value of the reaction coordinate which gives the fewest transition state levels at each energy.

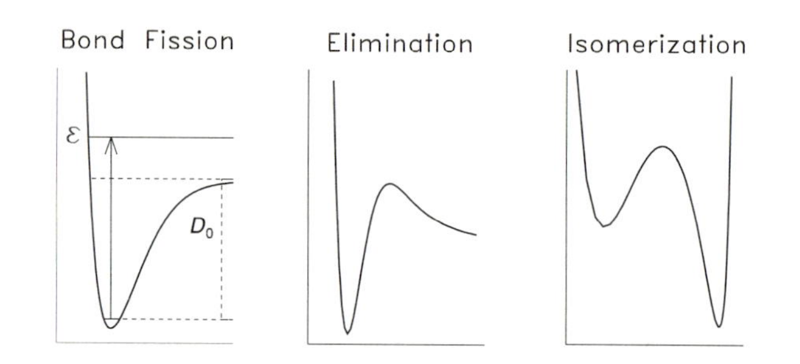

Fig. 5.12 Potential energy profiles for bond fission, elimination, and isomerization reactions.

Reactive trajectories at low energies have only just enough energy to cross the barrier, and therefore tend to follow the minimum energy path from reactants to products. As the energy is raised, trajectories begin to explore regions further from the minimum energy path, and are consequently more prone to be deflected back towards the reactants.

5.3 $k(\epsilon)$ for unimolecular reactions

A molecule undergoing a unimolecular reaction must first acquire sufficient energy to surmount the threshold energy to reaction, ϵ_0 (see Section 6.2). Here, we will focus on the unimolecular decay process of the energized molecule,

$$A^*(\epsilon) \longrightarrow \text{products},$$

and will demonstrate that the energy-specific rate coefficient for this process may be expressed in terms of cumulative reaction probabilities. Figure 5.12 illustrates the characteristic types of potential energy profile commonly encountered in unimolecular reactions, which are typical of bond fission, isomerization, and elimination reactions. The relevant energetics are also shown in the figure.

Relationship between $k(\epsilon)$ and $N(\epsilon)$.

The decay rate for a molecule with a specific energy, ϵ, is characterized by the microcanonical rate coefficient, $k(\epsilon)$. It is defined such that $k(\epsilon)$ represents the decay rate coefficient *per reactant state* for a molecule at energy ϵ. In terms of $k(\epsilon)$ the thermal rate coefficient can be written

For a thermally activated unimolecular reaction this expression applies to $k_\infty(T)$, the limiting high pressure rate coefficient: see Section 6.2.

$$k(T) = \frac{1}{q_r}\int_0^\infty k(\epsilon)\rho(\epsilon)\,e^{-\epsilon/k_B T}\,d\epsilon, \qquad (5.5)$$

where $\rho(\epsilon)\,d\epsilon$ is the number of reactant states in the energy range $\epsilon \rightarrow \epsilon + d\epsilon$, and $\rho(\epsilon)$ is referred to as the reactant *density of states* (with dimensions of 1/energy). Comparing the above equation with eqn 5.4 yields the following relationship between the microcanonical rate coefficient and $N(\epsilon)$:

$$k(\epsilon) = \frac{N(\epsilon)}{h\rho(\epsilon)}. \tag{5.6}$$

The right side of this expression has the correct units for a first order rate coefficient (s^{-1}).

A somewhat more refined expression for the microcanonical rate coefficient recognizes that the total angular momentum of the reactant molecule must be conserved during reaction:

$$k(\epsilon) = \sum_J k(\epsilon, J) = \sum_J \frac{N(\epsilon, J)}{h\rho(\epsilon, J)},$$

where $k(\epsilon, J)$ is the microcanonical rate coefficient at fixed energy, ϵ, and total angular momentum, J.

Evidence for the energy dependence of reaction rates

Unlike $N(\epsilon)$, the rate coefficient $k(\epsilon)$ can be measured directly, and early evidence for its dependence on energy was provided by chemical activation experiments. One such study looked at the reaction of H atoms with *cis*- and *trans*-but-2-ene and but-1-ene.[49] H atoms were generated by Hg-photosensitized decomposition of H_2. The three reactions each produced energized butyl radicals, which could either be stabilized by collisional relaxation with the excess H_2 bath gas present (the radicals thus formed can then react further to form butane as a major product), or dissociated to generate propene, for example

$$\text{H}+cis\text{-But-2-ene} \longrightarrow \text{Butyl}^* \xrightarrow{\text{M}} \text{Butyl}$$

$$\text{Butyl}^* \xrightarrow{k(\epsilon)} \text{CH}_3 + \text{Propene}.$$

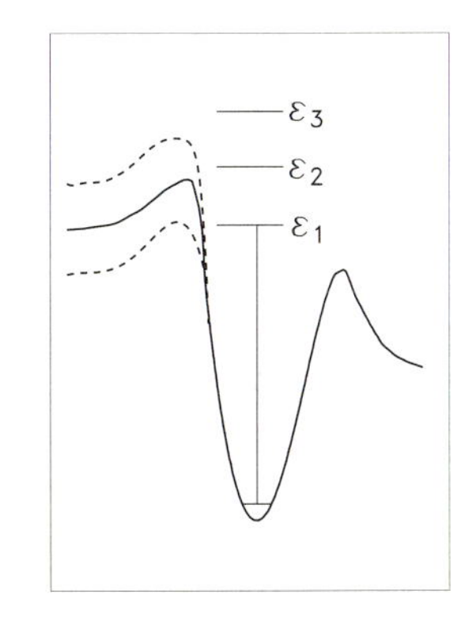

Fig. 5.13 Reaction profiles for the chemically activated production of butyl radicals.

The three different reactions produced energized butyl radicals with different internal energy contents (see Fig. 5.13), which could be controlled further by varying the temperature and the pressure. The reaction rate coefficients, determined from the relative yield of propene and butane, were found to depend on the source employed to generate the butyl radicals, and to *increase* with the radicals' internal energy. In a further series of experiments,[50] employing *cis*-but-2-ene, the reaction rate constant was measured over a very wide range of pressures, up to 20 MPa (200 atm). Varying the pressure altered the average lifetime of the energized molecule (i.e. the average time it took before the molecule underwent a collision with the bath gas, and was stabilized). At the highest pressures studied, the lifetime of the energized butyl radical was estimated to be as short as 2×10^{-13} s, and the reaction rate coefficients were found to be insensitive to lifetime in the range 1×10^{-8} to 2×10^{-13} s.

The energy dependence of the rate coefficients accorded well with RRKM theory (see below).

These experiments suggest not only that the rate coefficient increases with the internal energy in A*, but also that the timescale for reaction in chemically or thermally activated unimolecular reactions is typically considerably longer than that required for the energy to become randomized in the energized molecule. If it were not the case that energy is randomized rapidly, then the decay rate of the energized molecule would not be independent of the lifetime of the complex. Rapid energy randomization is the basis for the *free energy flow* approximation (see below), and is a fundamental assumption of many theories of unimolecular reactions.

In the absence of rapid energy randomization, the decay rate for molecules with excitation in different vibrational modes would vary. Such vibrational mode-specific behaviour would lead to rate coefficients which vary with lifetime, because at long times, for example, only those molecules with slow decay rates would survive.

Energy randomization was illustrated more strikingly in a later chemical activation study of the dissociation of the bicyclopropyl derivative generated by the following reaction,[51] in which the CH_2 reactants were generated by photolysis of ketene:

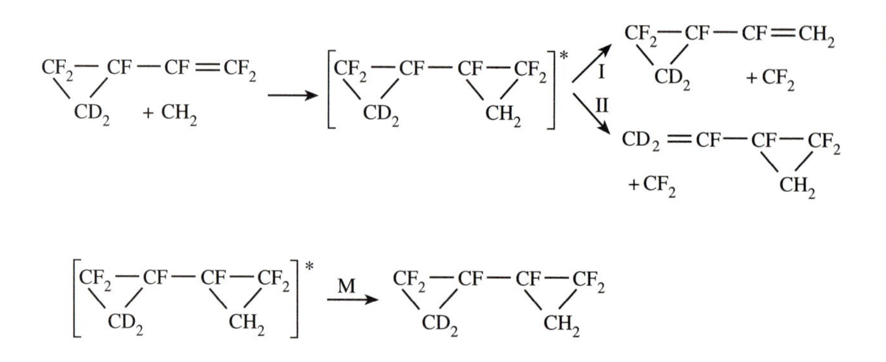

The products I and II were distinguished by their different fragmentation patterns in the mass spectrometer detector. The product ratio I/II was found to be close to unity and invariant of pressure in the range 0.8 to 310 torr (0.1–40 kPa). As the pressure was raised further, product I began to dominate. Thus, it was demonstrated that, at high pressures, when the energized intermediate survives for only a short time before it is stabilized, the vibrational energy in the energized intermediate has insufficient time to migrate from the ring initially excited in the chemical activation step. Detailed analysis suggested a timescale for energy redistribution between the two rings of $\sim 10^{-12}$ s.

Free energy flow and RRKM theory

Eqns 5.4 and 5.5 apply when the reactants are in thermal equilibrium at fixed temperature. Eqn 5.6, which was obtained by comparing the two expressions, also only applies under equilibrium conditions. $k(\epsilon)$ is a fixed energy, microcanonical rate coefficient, and eqn 5.6 is only valid when all reactant states, at a fixed total energy, are equally accessible. For a bimolecular reaction, in which the reactants are separated molecules or atoms, this constraint is a condition over which the experimentalist can exert some control; for example, by varying the pressure. In a unimolecular reaction, the reactant molecule is an energized molecule, and the condition that all states are equally accessible is not so easy for the experimentalist to ensure. The RRK and RRKM (Rice, Ramsperger, Kassel, and Marcus) theories of unimolecular reactions assume that, even if the internal states of the energized molecule are not initially equally populated, vibrational energy flows around the vibrational degrees of the molecule so rapidly, compared with the reaction rate, that the assumption of equally accessible internal states remains valid, i.e. they both assume rapid *intramolecular vibrational redistribution* (IVR).

Within the free energy flow assumption, eqn 5.6 yields the exact unimolecular rate coefficient at fixed energy. One can simplify the evaluation of the rate coefficient by making the further assumptions of transition state theory (assumptions 2 to 4 of Section 5.2). As in the bimolecular case, this allows the cumulative reaction probability, $N(\epsilon)$, to be replaced by the number of transition states below the energy ϵ, $N^{\ddagger}(\epsilon)$. The result is the RRKM expression for the microcanonical rate coefficient

As with eqn 5.6, eqn 5.7 for $k(\epsilon)$ is more rigorously written in terms of a sum over total angular momentum specific rate coefficients, $k(\epsilon, J)$.

$$k(\epsilon) = \frac{N^{\ddagger}(\epsilon)}{h\rho(\epsilon)}. \tag{5.7}$$

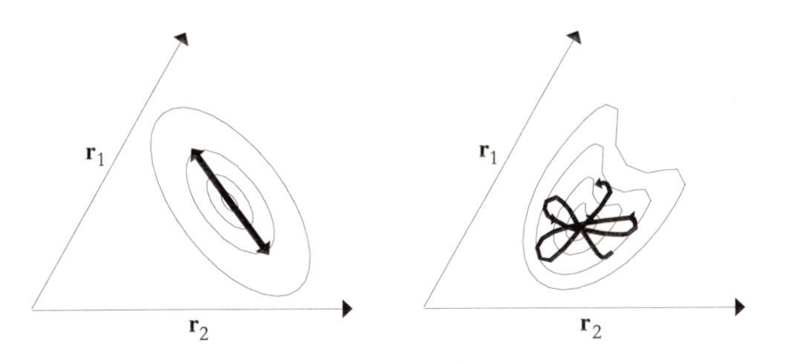

Fig. 5.14 Classical trajectories on an harmonic and an anharmonic potential surface.

Intramolecular vibrational redistribution

Much effort has been expended in recent years in establishing whether vibrational energy does indeed move freely about a molecule at high energies, and (if so) in determining the mechanism by which this redistribution takes place. Much of this work involves the spectroscopic interrogation of molecules very close to their reaction thresholds.[52] Although examples do exist of molecules whose vibrational energy remains highly localized, even up to very high energies, the assumption of rapid energy flow appears to be a remarkably robust one. A critical parameter is the density of reactant vibrational states, $\rho(\epsilon)$, which increases rapidly with molecular complexity (and hence with s, the number of vibrational modes in the molecule) and with total energy, ϵ. As a rule of thumb, if $\rho(\epsilon)$ exceeds ~ 50 states per cm^{-1}, the assumption of rapid free energy flow is likely to be valid, provided the reaction rate coefficient does not exceed $\sim 10^{12}$ s^{-1}. Thus, the free energy flow assumption of RRK and RRKM (and other, related 'statistical' models) is most likely to be valid for large molecules, with large reaction thresholds.

In Section 4.4 we saw that H_2O was one example of such a molecule.

Conversely, the free energy flow assumption breaks down if the reaction is prompt. This is often the case in photochemistry, in particular when molecular photodissociation occurs via a repulsive electronic state. Fragmentation then takes place on timescales of the order of the vibrational period $\ll 10^{-12}$ s, and the process is under the dynamical control of the potential energy surface.

There are many mechanisms by which vibrational energy can be redistributed in a molecule at high energies, all of which may be thought of as providing a means of coupling the 'normal' vibrational modes of the molecule together. The normal vibrational modes provide one (familiar) means of characterizing the vibrational motions (and spectra) of molecules at low energies. Molecular rotation is believed to play an important role in facilitating the coupling between these modes, by means of Coriolis and centrifugal forces. Anharmonicity in the potential energy surface is also a very important coupling mechanism. The effect of anharmonicity is illustrated in Fig. 5.14, in which schematic classical trajectories, representing vibrational motions on harmonic and anharmonic surfaces, are compared. In the harmonic case, trajectories are seen to oscillate for all time along the normal coordinates (the symmetric and antisymmetric stretch coordinates, say), while in the anharmonic case, trajectories are seen to explore the whole of the potential energy surface available at a given energy.

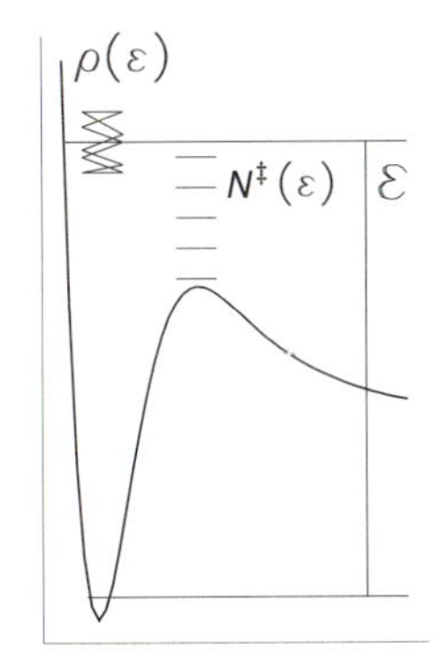

Fig. 5.15 Sums of states are evaluated at the transition state, densities of states are calculated for the energized molecule.

Densities and sums of states

Let us assume that the free energy flow assumption is valid, and return to the evaluation of $k(\epsilon)$ using the RRKM eqn 5.7. Figure 5.15 illustrates the species to which the sums and densities of states apply. The density of states is evaluated for the *activated* reactant molecule at energy ϵ, while the sum of

One of the reactant modes corresponds to the translational motion along the reaction coordinate, whereas the transition 'states' arise from motion orthogonal to the reaction coordinate.

states refers to the sum of internal states of the transition state species up to a total energy ϵ. At the transition state, the total energy is subdivided into the energy of the threshold, ϵ_0, defined as the zero-point energy difference between reactant and transition state, and an energy $\epsilon^\ddagger$ in the rovibrational modes of the TS. As is the case with bimolecular reactions, if the rovibrational levels of the transition state are known, then $N^\ddagger(\epsilon)$ may be evaluated by counting the number of transition states between the energies $\epsilon^\ddagger = 0$ to $\epsilon^\ddagger = \epsilon - \epsilon_0$, just as was done for the $H + H_2$ reaction in Section 5.2.

The density of *reactant* states at energy ϵ can be evaluated via a similar *direct count* procedure, by taking the derivative of the sum of *reactant* states $N_r(\epsilon)$, at energy ϵ, with respect to energy

$$\rho(\epsilon) = \frac{dN_r(\epsilon)}{d\epsilon},$$

where it is assumed that the level spacing is so small at high energies that the sum of states may be treated as a continuous function of ϵ. Efficient direct count algorithms are relatively easy to implement, and are widely used in the modelling of unimolecular reactions. However, given the uncertainty with which transition state and reactant vibrational frequencies are usually known at high energies, alternative classical and semi-classical methods for evaluating sums and densities of states have been developed, and are also widely employed.

The simplest of these methods assumes the vibrational modes to be harmonic. For s ($= 3N - 6$) *identical* harmonic oscillators of frequency ν, the total number of quanta to be distributed among these modes at energy ϵ is approximately $i = \epsilon/h\nu$. Each vibrational level, containing j quanta, has a degeneracy g_j, reflecting the number of equivalent ways of distributing j quanta in s identical modes:

These statistical mechanical expressions may look familiar: analogous expressions occur in Bose–Einstein statistics.

$$g_j = \frac{(j+s-1)!}{(j)!(s-1)!} = \frac{(j+s-1)\cdots(j+1)}{(s-1)!}.$$

If the vibrational level spacing is small, and there are a large number of quanta, j, to be distributed among the harmonic oscillator modes (i.e. $j \gg (s-1)$), then the degeneracies may be written

$$g_j \approx \frac{j^{s-1}}{(s-1)!}.$$

The total sum of states up to energy $\epsilon = ih\nu$ will, therefore, be the sum of (or, approximately, the integral over) all the degenerate levels

$$N_r(\epsilon) = \sum_{j=0}^{j=i} g_j \sim \frac{i^s}{s!},$$

or in terms of energies

$$N_r(\epsilon) = \frac{1}{s!}\left(\frac{\epsilon}{h\nu}\right)^s.$$

Applying this equation to the transition state species, which has an energy $\epsilon^\ddagger = \epsilon - \epsilon_0$ available to distribute in $s - 1$ internal vibrational modes, yields

$$N^\ddagger(\epsilon) = \frac{1}{(s-1)!}\left(\frac{\epsilon - \epsilon_0}{h\nu}\right)^{s-1}. \tag{5.8}$$

If the modes are not of equal frequency, then the sum of reactant states for s modes is given approximately by

$$N_r(\epsilon) = \frac{1}{s!}\left(\frac{\epsilon^s}{\Pi_k h\nu_k}\right),$$

where the product in the denominator is over *all* the s oscillators in the molecule.

The density of reactant states, for these harmonic oscillator cases, is the derivative of $N_r(\epsilon)$ with respect to energy. The resulting formula for s identical oscillators is

$$\rho(\epsilon) \sim \frac{1}{h\nu(s-1)!}\left(\frac{\epsilon}{h\nu}\right)^{s-1}, \qquad (5.9)$$

and, for inequivalent harmonic oscillators, is

$$\rho(\epsilon) = \frac{1}{(s-1)!}\left(\frac{\epsilon^{s-1}}{\Pi_k h\nu_k}\right).$$

More sophisticated expressions, which take into account the presence of zero-point energy and anharmonicity, can be found in texts given in the Background reading.

The effects of molecular rotation can be accounted for by evaluating the sums and densities of states as a function of total angular momentum quantum number, $\mathcal{J}$. This is achieved by adding to the potential energy the centrifugal kinetic energy (associated with the orbital motion of the separating products), and evaluating the sums and densities of states on the effective potential for each value of $\mathcal{J}$

RRK theory

In RRK theory a distinction is made between the energized molecule, $A_i^*(\epsilon)$, and the activated (transition state) molecule, $A_m^\ddagger(\epsilon)$:

$$A_i^*(\epsilon) \longrightarrow A_m^\ddagger(\epsilon) \xrightarrow{k^\ddagger} \text{products}$$

Any effects due to molecular rotation (see above) are ignored, and it is further assumed that

The derivation employed here makes use of the RRKM expression, eqn 5.7. Historically, RRKM theory was developed after RRK theory, and the original derivation by Rice and Ramsperger, later extended by Kassel, was rather different.

(1) there are s *harmonic* vibrational modes of the *same frequency*, ν, in the energized molecule, A^*, and that there are i quanta distributed in these s modes (i.e. $\epsilon = ih\nu$);
(2) in $A^\ddagger$, m quanta are localized in one vibrational mode, the reaction coordinate (i.e. $\epsilon_0 = mh\nu$), leaving $i-m$ quanta to be distributed in the $s-1$ remaining vibrational modes (these are again assumed to have the same frequency as in the energized molecule);
(3) energy can flow freely and rapidly in the energized and activated species: this allows the use of eqn 5.7 to calculate $k(\epsilon)$.

Substituting eqns 5.8 and 5.9 for the classical harmonic sums and densities of states into eqn 5.7 yields

$$k(\epsilon) = \frac{h\nu}{h}\left(\frac{\epsilon^\ddagger}{\epsilon}\right)^{s-1} = \nu\left(\frac{\epsilon-\epsilon_0}{\epsilon}\right)^{s-1}. \qquad (5.10)$$

In the original form of RRK theory, the ν pre-factor was instead given by $k^\ddagger$, the rate constant at which the activated molecule falls apart to form products. The theory did not provide a means by which this rate constant could be calculated, but it was assumed to be of the order of a vibrational frequency.

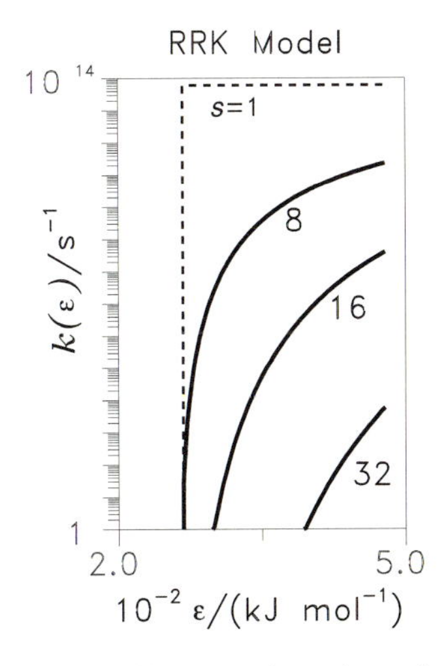

Fig. 5.16 The energy dependence of the RRK rate coefficient, $k(\epsilon)$ for different numbers of oscillators, s.

The energy dependence of $k(\epsilon)$ is illustrated in Fig. 5.16 for different numbers of oscillators s. As s increases (corresponding to increasing molecular complexity), the rate coefficients decrease. The 'statistical' interpretation of this finding is clear: as the number of vibrational modes increases, there is less chance that energy ϵ_0, out of the total ϵ, will reside in the reaction coordinate. When there is only one vibrational mode in the molecule, the probability that the reaction coordinate contains at least the energy ϵ_0 is unity, provided $\epsilon \geq \epsilon_0$.

5.4 The measurement of $k(\epsilon)$

Although femtosecond laser pulses offer superior time resolution, their energy resolution is inferior to picosecond pulses, as a consequence of the energy–time 'uncertainty' relationship; see Atkins and Friedman's *Molecular quantum mechanics*.

Many methods have been developed to measure $k(\epsilon)$. The most direct of these make use of pico- and femtosecond pulsed laser radiation. One such method uses the photochemical process of internal conversion (IC), a collisionless, non-radiative event, in which a molecule crosses from one excited electronic state to another of the same electron spin multiplicity. The method has been applied to the dissociation of ketene.[53] Pulsed laser radiation was used to excite electronically the ketene molecule to its first excited singlet state (S_1), from which it undergoes very rapid IC to the ground electronic state (S_0):

$$CH_2CO \xrightarrow{h\nu} CH_2CO(S_1) \xrightarrow{IC} CH_2CO^*(S_0).$$

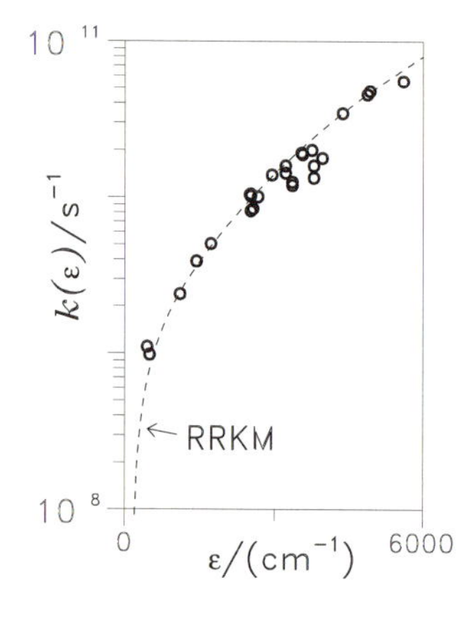

Fig. 5.17 Experimental $k(\epsilon)$ values for ketene dissociation compared with RRKM calculated rate coefficients.

The internal conversion process conserves total energy, and thus the ground state molecule is produced in very high rovibrational levels, which lie above the threshold energy for dissociation,

$$CH_2CO^*(S_0) \xrightarrow{k(\epsilon)} CH_2 + CO.$$

The CH_2 fragment was detected by time-resolved laser induced fluorescence (see Section 4.1). The dissociation step in the above sequence was believed to be rate determining, and the reaction rate coefficients, $k(\epsilon)$, were obtained by monitoring the CH_2 fluorescence intensity as a function of delay time between the 'pump' excitation laser pulse and the 'probe' LIF laser pulse. The results are summarized in Fig. 5.17, where they are compared with detailed RRKM calculation. Further experiments on this system, reviewed elsewhere,[41, 52] are believed to demonstrate the appearance of staircase structure in $k(\epsilon)$, reflecting the staircase structure already noted in $N(\epsilon)$.[54]

An alternative method employs direct laser excitation to high-energy, quasi-bound vibrational overtone levels. The transitions with the highest intensities are generally those involving excitation to X–H stretching overtone levels, since the latter have high anharmonicities. Real time laser pump–probe experiments, analogous to that described above, have been performed, but many studies have made use of the same collisional relaxation 'clock' that was employed in the pioneering chemical activation studies discussed earlier. One example is the study of the isomerization of cyclobutene to butadiene, in which the excitation was initially deposited in methylenic and olefinic C–H stretching overtone levels in the energy region of $6\nu_{CH}$.[55] The data are summarized in Fig. 5.18, where they are compared with the simpler RRK model (see Section 5.3). To fit the experimental $k(\epsilon)$ data using the classical RRK expression (equation 5.10) requires $s \sim 10$; i.e. approximately half the number of oscillators in the molecule ($3N - 6$). This might be taken as suggesting that only half the oscillators in the molecule participate in energy

To apply the RRK model, $k^{\ddagger}$ was set equal to the limiting high pressure *A*-factor, $A_\infty = 5.75 \times 10^{13}\ s^{-1}$ (see Section 6.2). The threshold energy ϵ_0 for the reaction is 11600 cm^{-1}.

flow. However, this is not the case. The RRK assumptions, that all the oscillators in the molecule and the TS are identical and harmonic, are extreme ones. Much better agreement with experiment is obtained assuming that *all* modes participate in IVR, provided that the RRKM expression for $k(\epsilon)$ is evaluated using direct count procedures for the sums and densities of states (or more reliable semi-classical approximations thereof), and proper account is taken of the different vibrational frequencies of the TS and reactant modes. This conclusion is reinforced by the data shown in Fig. 5.18, which reveal that the reaction rate depends only on the energy of the energized molecule, and not on the type of vibrational mode initially excited.

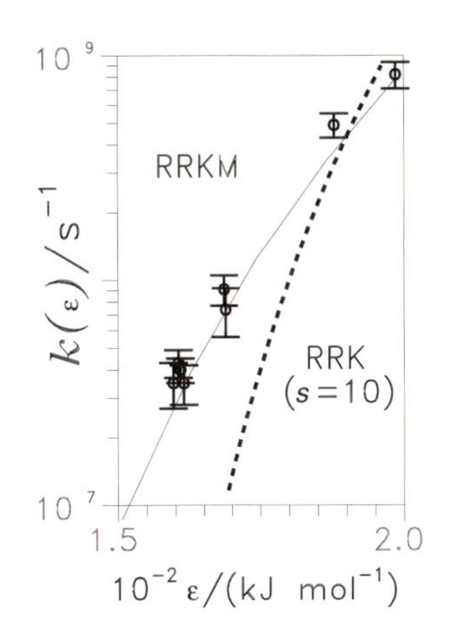

Fig. 5.18 Rate coefficients, $k(\epsilon)$, for cyclobutene isomerization compared with RRK theory (with $s = 10$) and RRKM theory.

6 Thermal rate coefficients

6.1 Canonical transition state theory (CTST)

In the following section we will focus on the derivation and application of transition state theory to bimolecular reactions.

Derivation

Consider, once more, the triatomic reaction

$$A + BC \longrightarrow ABC^{\ddagger} \longrightarrow AB + C.$$

We have seen in Section 5.2 that TST leads to the following expression for the transition state reaction probabilities,

$$P_n(\epsilon) = 1 \qquad \text{if } \epsilon > \epsilon_n^{\ddagger} + \epsilon_0.$$

Substitution into eqn 5.3, with a change of variable of integration to $\epsilon_t^{\ddagger}$, yields

$$\begin{aligned} k(T) &= \frac{1}{h\, q_t} \int_0^{\infty} \frac{\sum_n g_n^{\ddagger}\, e^{-\epsilon_n^{\ddagger}/k_B T}}{q_{A:int}\, q_{BC:int}}\, e^{-\epsilon_0/k_B T}\, e^{-\epsilon_t^{\ddagger}/k_B T}\, d\epsilon_t^{\ddagger} \\ &= \frac{\sum_n g_n^{\ddagger}\, e^{-\epsilon_n^{\ddagger}/k_B T}}{h\, q_t\, q_{A:int}\, q_{BC:int}}\, e^{-\epsilon_0/k_B T} \int_0^{\infty} e^{-\epsilon_t^{\ddagger}/k_B T}\, d\epsilon_t^{\ddagger} \end{aligned}$$

Evaluating the integral over translational energies leads to the following TS expression for the thermal rate coefficient:

$$k(T) = \frac{k_B T}{h} \frac{q_{int}^{\ddagger}}{q_t\, q_{A:int}\, q_{BC:int}}\, e^{-\epsilon_0/k_B T}, \tag{6.1}$$

where

$q_{int}^{\ddagger}$ *excludes* the translational degree of freedom corresponding to motion along the reaction coordinate.

$$q_{int}^{\ddagger} = \sum_n g_n^{\ddagger}\, e^{-\epsilon_n^{\ddagger}/k_B T} = q_{rot}^{\ddagger}\, q_{vib}^{\ddagger}\, q_{el}^{\ddagger},$$

$$q_{BC:int} = q_{BC:rot}\, q_{BC:vib}\, q_{BC:el},$$

and (since $\mu = m_A m_{BC}/M$, where M is the total mass)

$$q_t = \left(\frac{2\pi\mu k_B T}{h^2}\right)^{3/2} \equiv \left(\frac{q_t^{\ddagger}}{q_{A:t}\, q_{BC:t}}\right)^{-1}.$$

Eqn 6.1 can thus be rewritten, more familiarly, in terms of the total molecular partition functions of the transition state and reactant molecules, as

CTST may also be cast in *thermodynamic* form (see Benson's *Thermochemical kinetics*).

$$k(T) = \frac{k_B T}{h} \frac{q^{\ddagger}}{q_A\, q_{BC}}\, e^{-\epsilon_0/k_B T}. \tag{6.2}$$

Applications of canonical transition state theory

Recombination of two atoms: simple collision theory revisited. Imagine the atoms as structureless hard spheres which react if the reactant collision energy along the line-of-centres exceeds the barrier height:

$$\mathrm{A} + \mathrm{B} \longrightarrow \mathrm{AB}^{\ddagger} \longrightarrow \text{products}.$$

Stable AB products could only be formed if the excited AB species produced on atom combination were stabilized by collisions with a high pressure of an added bath gas.

The transition state partition function is

$$q^{\ddagger}_{\mathrm{int}} = q^{\ddagger}_{\mathrm{rot}} = \frac{k_{\mathrm{B}}T}{B^{\ddagger}},$$

where the single 'vibrational' mode of $\mathrm{AB}^{\ddagger}$ is the reaction coordinate, and hence is *not* included in the evaluation of the transition state partition function. The transition state rotational constant,

$$B^{\ddagger} = \frac{h^2}{8\pi^2 \mu d^2},$$

where

$$d = r_{\mathrm{A}} + r_{\mathrm{B}},$$

is evaluated on the assumption that the transition state corresponds to the particles touching. The translational partition function is

$$q_{\mathrm{t}} = \left(\frac{2\pi\mu k_{\mathrm{B}}T}{h^2}\right)^{3/2} \equiv \left(\frac{q^{\ddagger}_{\mathrm{t}}}{q_{\mathrm{A:t}}\, q_{\mathrm{BC:t}}}\right)^{-1},$$

and, for the internal degrees of freedom of the structureless reactants,

$$q_{\mathrm{A:int}} = q_{\mathrm{B:int}} = 1.$$

Substitution into the expression for $k(T)$, eqn 6.1, yields

$$k(T) = \left(\frac{8k_{\mathrm{B}}T}{\pi\mu}\right)^{1/2} \pi d^2\, \mathrm{e}^{-\epsilon_0/k_{\mathrm{B}}T} \qquad (6.3)$$

which, perhaps surprisingly, is the same expression as derived from simple collision theory. Note the recovery of the collision cross-section, πd^2.

The steric factor. A simplified TST expression may be used to obtain an order of magnitude estimate for the steric factor, P, of a reaction, which is not available *a priori* from simple collision theory (SCT). First, the SCT expression for the rate coefficient can be written in terms of TST,

$$k_{\mathrm{SCT}}(T) \sim \frac{k_{\mathrm{B}}T}{h}\frac{q^2_{\mathrm{rot}}}{q^3_{\mathrm{t}}}\,\mathrm{e}^{-\epsilon_0/k_{\mathrm{B}}T},$$

where all the partition functions have been redefined in terms of partition functions per degree of freedom. Take the specific example of an AB + CD reaction proceeding via a non-linear TS. The rate coefficient can be expressed as

$$k_{\mathrm{TST}}(T) \sim \frac{k_{\mathrm{B}}T}{h}\frac{q^3_{\mathrm{rot}}q^5_{\mathrm{vib}}}{q^3_{\mathrm{t}}q^4_{\mathrm{rot}}q^2_{\mathrm{vib}}}\,\mathrm{e}^{-\epsilon_0/k_{\mathrm{B}}T}.$$

Taking the ratio of the TST and the SCT expressions yields the steric factor

$$P = \frac{k_{\mathrm{TST}}(T)}{k_{\mathrm{SCT}}(T)} \sim \left(\frac{q_{\mathrm{vib}}}{q_{\mathrm{rot}}}\right)^3 \sim 10^{-2} - 10^{-4},$$

Table 6.1 Expressions for the partition functions employed in CTST for linear and non-linear molecules

Motion	Symbol	Degrees of freedom	Partition function	Order of magnitude
Translation	$q_{A:t}$	3	$\left(\frac{2\pi m_A k_B T}{h^2}\right)^{3/2}$	$10^{30}-10^{33}$ m^{-3}
Rotation (linear)	$q_{BC:rot}$	2	$\frac{k_B T}{\sigma hcB}$	10^1-10^2
Rotation (non-linear)	$q_{ABC:rot}$	3	$\frac{1}{\sigma}\left(\frac{k_B T}{hc}\right)^{3/2}\left(\frac{\pi}{ABC}\right)^{1/2}$	10^2-10^3
Vibration (linear)	$q_{ABC:vib}$	$3N-5$	$\prod_i^{3N-5} q^i_{vib}$	$1-10$
Vibration (non-linear)	$q_{ABC:vib}$	$3N-6$	$\prod_i^{3N-6} q^i_{vib}$	$1-10$
TS vibration (linear)	$q^{\ddagger}_{vib}$	$3N-6$	$\prod_i^{3N-6} q^i_{vib}$	$1-10$
TS vibration (non-linear)	$q^{\ddagger}_{vib}$	$3N-7$	$\prod_i^{3N-7} q^i_{vib}$	$1-10$
Electronic	$q_{BC:el}$	—	$\sum_i g_i e^{-\epsilon_i/k_B T}$	g_0

$q^i_{vib} = \left(1 - e^{-hc\nu_i/k_B T}\right)^{-1}$.
Rotational constants, A, B, and C, and vibrational frequencies, ν, expressed in cm^{-1}.

where use has been made of the order of magnitude estimates of the partition functions (at room temperature) given in Table 6.1. Analogous procedures suggest that, in general, as the reactant complexity increases, the A-factor predicted by TST falls, as does the estimated steric factor.

The $H + D_2$ reaction. We now turn to a quantitative assessment of canonical transition state theory, and calculate the thermal rate coefficient for the reaction

$$\mathrm{H + D_2 \longrightarrow HD + D.}$$

To evaluate eqn 6.1, one needs to know the geometry and vibrational frequencies of the transition state molecule. Both quantities are obtained from the potential energy surface: the latter has been calculated with high precision for the $H + H_2$ system, and the data are summarized in Table 6.2. The geometry of the transition state (i.e. at the barrier) is linear. At 300 K, we may evaluate each term as follows (where ϵ_z refers to the zero-point energy):

$$\begin{aligned}
\epsilon_0 &= V^{\ddagger} + \epsilon_z^{\mathrm{HDD}} - \epsilon_z^{\mathrm{DD}} \\
&= 39.91 + 18.84 - 18.59 = 40.16 \text{ kJ mol}^{-1} \\
e^{-\epsilon_0/k_B T} &= 1.02 \times 10^{-7} \\
\frac{k_B T}{h} &= 6.25 \times 10^{12} \text{ s}^{-1} \\
q_t &= \left(\frac{2\pi\mu k_B T}{h^2}\right)^{3/2} = 6.99 \times 10^{29} \text{ m}^{-3}.
\end{aligned}$$

Table 6.2 TS parameters for H + D_2

Parameter	Reactants (D_2)	Transition state
$r_{e,\,HD}$/Å	—	0.93
$r_{e,\,DD}$/Å	0.741	0.93
Potential energy/kJ mol^{-1}	0.0	39.91
Frequencies/cm^{-1}		
Stretch	3109	1762
Bend (doubly degenerate)	—	694

For the transition state,

$$I^{\ddagger} = \frac{m_1 m_2 + 4m_1 m_3 + m_2 m_3}{M} r_e^{\ddagger 2} = 4.02 \times 10^{-47} \text{ kg m}^2$$

$$B^{\ddagger} = \frac{\hbar^2}{2I^{\ddagger}} = 1.38 \times 10^{-22} \text{ J molecule}^{-1}$$

$$q^{\ddagger}_{\text{rot}} = \frac{k_B T}{B^{\ddagger}} = 30.0$$

$$q^{\ddagger}_{\text{vib}} = \prod_i \left(1 - e^{-h\nu_i/k_B T}\right)^{-1} = 1.0002 \times (1.037)^2 = 1.076.$$

For the reactants,

We are assuming that the high temperature expressions for the rotational partition functions can be employed. Even at 300 K little error is introduced by so doing.

$$I_{D_2} = 9.14 \times 10^{-48} \text{ kg m}^2$$

$$B_{D_2} = \frac{\hbar^2}{2I_{D_2}} = 6.08 \times 10^{-22} \text{ J molecule}^{-1}$$

$$q_{D_2:\text{rot}} = \frac{k_B T}{2B} = 3.41$$

$$q_{D_2:\text{vib}} = \left(1 - e^{-h\nu/k_B T}\right)^{-1} = 1.00.$$

Substitution into the expression for $k(T)$ yields

$$\begin{aligned} k(T) &= 8.5 \times 10^{-17} e^{-\epsilon_0/k_B T} \\ &= 8.6 \times 10^{-24} \text{ m}^3 \text{ molecule}^{-1} \text{ s}^{-1} \\ &\equiv 8.6 \times 10^{-18} \text{ cm}^3 \text{ molecule}^{-1} \text{ s}^{-1} \end{aligned}$$

The experimental[56] and the quantum mechanically calculated[57] rate coefficients at 300 K are 2.1×10^{-17} cm^3 molecule^{-1} s^{-1} and 1.9×10^{-17} cm^3 molecule^{-1} s^{-1}. The quasi-classical trajectory calculated value[58] is 1.8×10^{-17} cm^3 molecule^{-1} s^{-1}. The results are shown more fully in the form of an Arrhenius plot in Fig. 6.1.

The discrepancy between CTST and the experimental (or QM) rate coefficients can be ascribed to neglect of tunnelling in the CTST calculation (see below). Note, however, that classical mechanics works rather well, suggesting that the absence of a tunnelling contribution to the classical rate coefficient is offset by the absence of zero point vibrational energy in the barrier region, which lowers the effective barrier height compared with the quantum mechanical one. The CTST calculation treats motion along the reaction coordinate classically (i.e. there is no tunnelling). However,

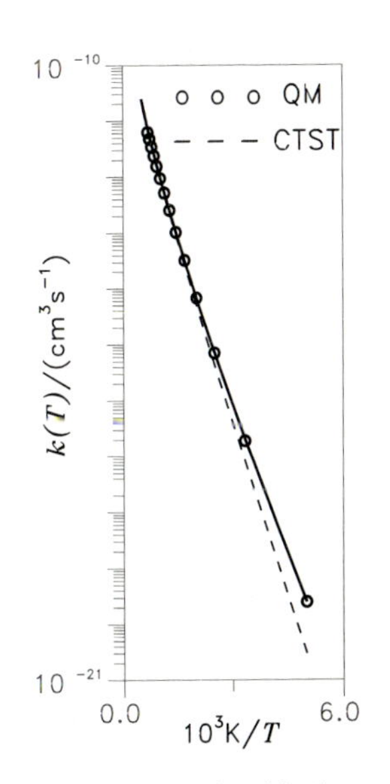

Fig. 6.1 Arrhenius plot of the thermal rate coefficient for H + D_2 reaction.

A fuller explanation for the good agreement between the QCT and QM $k(T)$ can be found elsewhere.[12]

zero-point energy at the transition state *is* included (because motion orthogonal to the MEP is treated quantum mechanically in this version of CTST). The latter leads to a higher barrier than is the case in the classical treatment. Thus, in this sense, CTST represents the worst of both worlds (a high barrier with no tunnelling) and leads to higher activation energies and lower rate coefficients than are observed experimentally.

These corrections are still approximate because the true quantum mechanical motion is not separable; see Section 5.2. In fact, there is no rigorous quantum mechanical version of TST.[1]

Tunnelling corrections. Significant improvements to transition state theory can be achieved by including approximate, one-dimensional tunnelling corrections to account for the quantum mechanical motion through the reaction barrier. The approximate correction to the thermal (canonical) TST rate coefficient, κ, is given as an integral over the translation energy of the one-dimensional barrier transmission probability:

$$\kappa = \int_{-\infty}^{\infty} T(\epsilon_t)\, e^{-\epsilon_t/k_B T} d\epsilon_t.$$

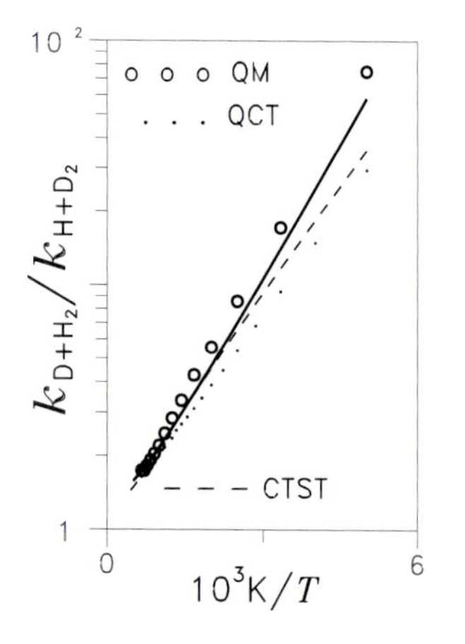

Fig. 6.2 The temperature dependence of the isotope effect (k_{D+H_2}/k_{H+D_2}) for the hydrogen exchange reactions.

$T(\epsilon_t)$ can be evaluated approximately, assuming the barrier to be either of parabolic form (i.e. using eqn 5.2, setting the barrier height to zero), or, more realistically, using the Eckart barrier. The factor $e^{-\epsilon_t/k_B T}$ in the integral ensures that tunnelling corrections become more important at low temperature, as is necessary to account for the data shown in Fig. 6.1.

Isotope effects. Transition state theory provides an excellent means of determining isotope effects, because errors in the transition state properties (frequencies and geometry) tend to cancel out when the ratio of rate coefficients for two isotopic reactions is taken. In Arrhenius form, the rate coefficient ratio for a reaction involving an H atom and a D atom may be written

$$\ln\frac{k_H}{k_D} = \ln\frac{A_H}{A_D} - \frac{E_{aH} - E_{aD}}{RT},$$

where $E_{a\,D,H}$ are the activation energies of the reactions with and without isotopic substitution. The dominant term in this expression at low temperatures is the second term, arising from the difference in activation energies for the two reactions.

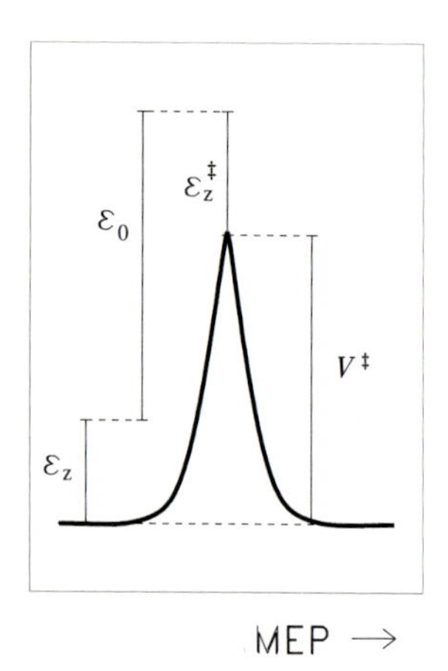

Fig. 6.3 Energies employed in making a CTST estimate of the isotope effect.

Figure 6.2 illustrates the kinetic isotope effects for the reactions $D + H_2$ and $H + D_2$.[56] The gradient of the line in this diagram is approximately determined by the difference in activation energies for the two reactions, which in turn is dominated by the change in the reactant and TS zero-point energies upon isotopic substitution (see Fig. 6.3 and the following subsection). The intercept of the graph is the ratio of A-factors in the limit of high temperature, which, in transition state theory, is given by a ratio of partition functions. For this reaction, the ratio of the A-factors is close to unity at high temperatures. Ironically, the CTST predicted isotope effect is actually closer to the experimental data than that obtained via the QCT method, even though the latter yields much better agreement with the absolute experimental rate coefficients.

An upper bound to the kinetic isotope effect may be obtained assuming that the zero-point energies of the transition state species are insensitive to isotopic substitution. This is a reasonable assumption for reactions in which the bond involving deuterium substitution is significantly weakened in the transition state, in which case the difference in activation energies may be approximated by the zero-point energy differences of the reactants,

$$E_{\mathrm{aH}} - E_{\mathrm{aD}} \sim E_{\mathrm{zD}} - E_{\mathrm{zH}}.$$

The relevant energetics are summarized in Fig. 6.3.

Temperature dependences. A qualitative idea about the temperature dependence of the rate coefficient predicted by TST can be obtained by assuming the vibrational partition functions are unity. Consider an A + BC reaction proceeding via a linear transition state. In this case, the predicted temperature dependence can be extracted from the various partition functions in eqn 6.2 as

$$\begin{aligned} k(T) &= C\,T^1 \frac{T^1}{T^{\frac{3}{2}}\,T^1}\,\mathrm{e}^{-E_0/RT} \\ &= C\,T^{-\frac{1}{2}}\mathrm{e}^{-E_0/RT}, \end{aligned}$$

where C is a constant. In general, the temperature dependence of the bimolecular rate coefficient has the form

$$k(T) = C\,T^n\,\mathrm{e}^{-E_0/RT},$$

and from the definition of the activation energy, eqn 1.2, one obtains

$$\frac{\mathrm{d}\ln k(T)}{\mathrm{d}T} = \frac{nRT + E_0}{RT^2} = \frac{E_\mathrm{a}}{RT^2},$$

i.e.

$$E_\mathrm{a} = E_0 + nRT.$$

Provided that $E_0 \gg RT$, the temperature dependence of the rate coefficient is thus dominated by the zero-point energy difference between the reactants and transition state, E_0.

Tunnelling through the reaction barrier has already been shown to be one cause of curvature in the Arrhenius plot, which becomes particularly apparent at low temperatures for reactions involving light species. As shown above another source of curvature is the temperature dependence of the partition functions appearing in the TST expression for the *A*-factor, which will be most evident for reactions with low activation barriers. Particularly striking examples include the OH + H_2 reaction (see below), and the reaction

$$\mathrm{OH} + \mathrm{CO} \longrightarrow \mathrm{HOCO}^* \longrightarrow \mathrm{H} + \mathrm{CO_2},$$

which proceeds via an energized HOCO complex (see Section 3.3). An Arrhenius plot of the rate coefficient data for this reaction is shown in Fig. 6.4.[59] The reaction is best modelled as an association reaction, followed by dissociation of the HOCO intermediate.[60] The unusual temperature dependence is thought to arise from the competition between dissociation of the HOCO complex back to reactants, and dissociation to form products. The rate of the latter is believed to increase rapidly with temperature, partly because the vibrational partition functions for the low frequency vibrational modes of the second transition state increase rapidly with temperature. Tunnelling through the second barrier may also be important.[41,61]

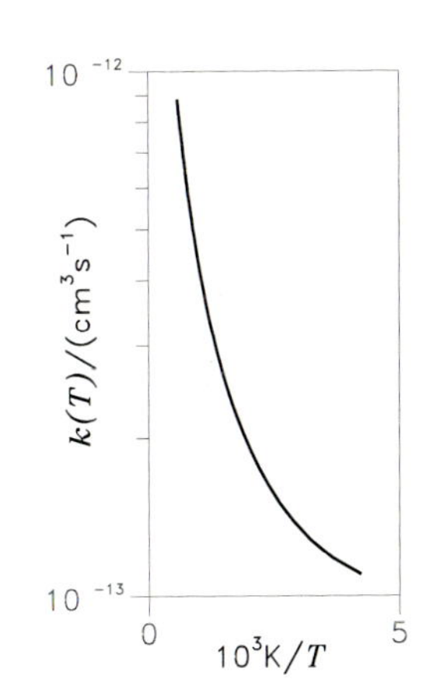

Fig. 6.4 Arrhenius plot of the rate coefficient for the OH + CO reaction.

Non-Arrhenius behaviour may also arise if *initial state specific* rate coefficients display a marked dependence on reactant quantum state. This could be an enhancement (or reduction) of state specific rate coefficient with reactant rotational or vibrational excitation. An example of curvature in the Arrhenius plot arising from vibrational enhancement of the state specific rate coefficients is the reaction (see Section 4.4)

$$\mathrm{OH} + \mathrm{H_2}(v = 0, 1) \longrightarrow \mathrm{H_2O} + \mathrm{H},$$

The data are discussed in more detail by Pilling and Smith in *Modern gas kinetics*.

where the rate enhancement on vibrational excitation of H_2 is over a factor of $\times 100$ at room temperature. The state-specific rate coefficients have been measured as a function of temperature, and although they appear to fit the Arrhenius equation (albeit over a rather narrow temperature range), an Arrhenius plot of the *thermal* rate coefficient is predicted to display considerable curvature at high temperatures, where $H_2(v = 1)$ becomes significantly populated.

The curved Arrhenius behaviour for this reaction is reasonably well accounted for by CTST, provided a tunnelling correction is made. The experimental data are shown in Fig. 6.5,[62] where they are compared with the results of precise QM calculations,[63] and with the predictions of CTST without a 1-D tunnelling correction. Both calculations employ the same *ab initio* potential energy surface.[64] The more exact QM results clearly indicate that the surface employed is defective. The QM calculated rate coefficients are significantly larger than the experimental ones, particularly at low temperature, suggesting that the importance of tunnelling is overestimated. More recent *ab initio* potential energy surface calculations[2] support the view that the width of the barrier on the PES employed in the above calculations is too narrow.

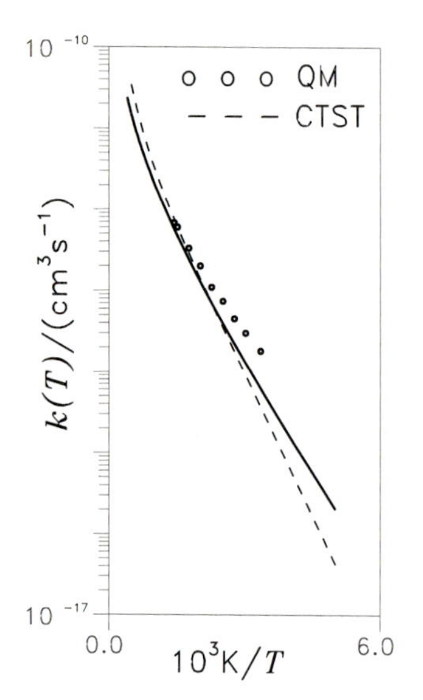

Fig. 6.5 Arrhenius plot of the rate coefficient for the $OH + H_2$ reaction.

6.2 Thermally activated unimolecular reactions

The energy dependent Lindemann scheme

We have seen (in Section 5.3) that the rate coefficient for a molecule undergoing unimolecular reaction depends on energy. In a thermally activated unimolecular reaction, the molecule acquires the energy necessary to react by energy-transferring collisions, and like the elementary reaction step, the rate coefficients of these activating and deactivating collisions also depend on the energy of the energized molecule. Consider the modified, energy dependent Lindemann scheme, involving an energized molecule, $A^*(\epsilon)$, with a specific internal energy ϵ, which exceeds the threshold energy to reaction ϵ_0. The elementary activation, deactivation, and reactive processes may be written:

The energy independent Lindemann scheme, which is the basis of *Lindemann theory*, is discussed in many of the texts given in the Background reading.

$$\begin{aligned}
\mathrm{A} + \mathrm{M} &\xrightarrow{k_1(\epsilon)} \mathrm{A}^*(\epsilon) + \mathrm{M} \\
\mathrm{A}^*(\epsilon) + \mathrm{M} &\xrightarrow{k_{-1}(\epsilon)} \mathrm{A} + \mathrm{M} \\
\mathrm{A}^*(\epsilon) &\xrightarrow{k(\epsilon)} \text{products(P)}.
\end{aligned}$$

M refers to a *bath gas*, which could be the molecule itself, but is more usually an inert or unreactive gas. Its concentration is generally much greater than those of the other reactants, and thus [M] defines the pressure of the experiment, via application of the gas laws. The energy-specific rate of product formation may be written

$$\frac{d[P]}{dt} = k_{uni}(\epsilon)[A] = k(\epsilon)[A^*(\epsilon)].$$

Applying the steady state approximation to the concentration of the energy-specific energized molecule, $A^*(\epsilon)$ yields the energy-specific unimolecular rate coefficient

$$k_{uni}(\epsilon) = \frac{[A^*(\epsilon)]}{[A]}k(\epsilon) = \frac{k_1(\epsilon)[M]}{k_{-1}(\epsilon)[M] + k(\epsilon)}k(\epsilon).$$

In the steady state approximation, the rate of change of concentration of the energized molecule, $d[A^*(\epsilon)]/dt$, is set equal to zero. This approach is strictly only valid in the limit $k_1(\epsilon)[M] \ll k_{-1}(\epsilon)[M] + k(\epsilon)$, but comparison with the exact solution shows this to be a reasonable approximation provided the critical energy $\epsilon_0 \gg k_B T$.

Integrating over energies, we obtain the unimolecular rate coefficient,

$$k_{uni}(T) = \int_{\epsilon_0}^{\infty} n(\epsilon)\, k(\epsilon)\, d\epsilon, \qquad (6.4)$$

where $k(\epsilon)$ is the energy-dependent reaction rate coefficient (see Section 5.3), and

$$n(\epsilon)d\epsilon = \frac{k_1(\epsilon)\,[M]}{k_{-1}(\epsilon)[M] + k(\epsilon)}\, d\epsilon,$$

represents the fraction (or population) of A molecules in the energy range $\epsilon \to \epsilon + d\epsilon$. Moreover, the product $n(\epsilon)\, k(\epsilon)\, d\epsilon$ in eqn 6.4 represents the contribution to the unimolecular reaction rate coefficient in this energy range; i.e. $k_{uni}(\epsilon)\, d\epsilon$.

The strong collision assumption. Implicit in the modified Lindemann scheme is the assumption that deactivating collisions remove sufficient energy from $A^*(\epsilon)$ to render it unreactive. If this is so, $k_{-1}(\epsilon)$ will equal the collision number, Z^0_{AM}, irrespective of energy. Single collisions which relax $A^*(\epsilon)$ to A are usually termed *strong collisions*.

More detailed calculations suggest that a somewhat better approximation is to assume that only a fraction, λ, of collisions is capable of deactivating $A^*(\epsilon)$ to A, i.e.

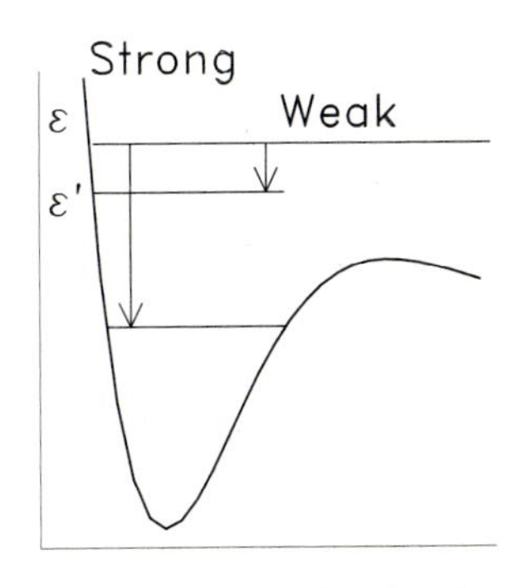

Fig. 6.6 Comparison of weak and strong collisions.

$$k_{-1}(\epsilon) = \lambda\, Z^0_{AM} = \frac{\omega}{[M]}.$$

λ is usually treated as an adjustable parameter used to fit the fall-off curve (the variation in k_{uni} with pressure). It takes approximate account of *weak collisions* which partially deactivate $A^*(\epsilon)$ to energies (ϵ') which still lie above the threshold energy. Weak collisions are illustrated in Fig. 6.6.

Within this approximation, the populations $n(\epsilon)$, eqn 6.4, may be rewritten in terms of ω:

$$n(\epsilon)\, d\epsilon = \frac{k_1(\epsilon)\,[M]}{\omega + k(\epsilon)}\, d\epsilon. \qquad (6.5)$$

Other treatments have been proposed to take better account of weak collisions. One approach involves solving the *Master equation*, which describes the coupled rate equations for the energized molecules with different energy contents. This approach is described in detail by Robinson and Holbrook in *Unimolecular reactions*.

Pressure-dependent populations. At very high pressures, the activation and deactivation rates are equal, and greatly exceed the reaction rate, and A^* and A are in thermal equilibrium. One may write, with the help of eqn 6.4,

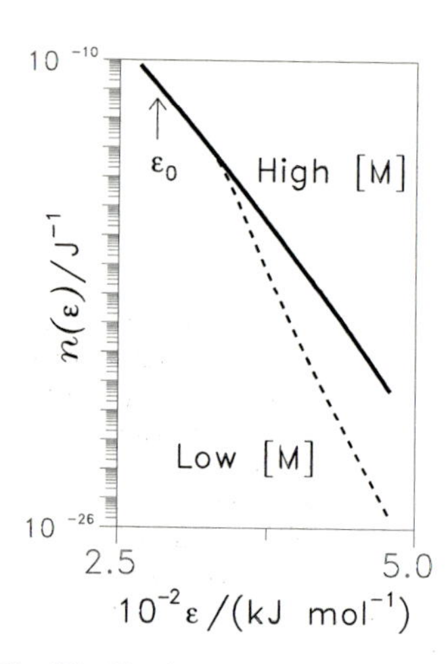

Fig. 6.7 The dependence of the populations $n(\epsilon)$ on the pressure.

$$n(\epsilon) = \frac{[A^*(\epsilon)]}{[A]} = \frac{k_1(\epsilon)}{k_{-1}(\epsilon)} \qquad [M] \to \infty$$
$$= \frac{\rho(\epsilon)\,e^{-\epsilon/k_B T}}{q_A}, \tag{6.6}$$

where the latter equation is the Boltzmann expression for the fraction of A molecules at energy ϵ at *thermal equilibrium*. In the continuous energy limit, state degeneracies, g, are replaced by the energy dependent reactant density of states, $\rho(\epsilon)$, introduced in Section 5.3. Eqn 6.6 demonstrates the necessity of considering the energy dependence of the activation and deactivation processes. Within the strong collision assumption, the rate coefficient for the latter is energy *in*dependent, and thus $k_1(\epsilon)$ must be energy dependent in order to satisfy the equilibrium condition of eqn 6.6.

The energy-dependent populations are illustrated in Fig. 6.7. In the high pressure limit, the population distribution is given by the Boltzmann law. However, as the pressure is decreased, the equilibrium between A^* and A is perturbed by the reaction, and the distribution of the energized molecules is no longer given by eqn 6.6. The most highly excited A^* molecules are preferentially depleted by reaction, since (as we have seen) they have the fastest reaction rates. As a consequence, the average energy of those molecules undergoing reaction varies with pressure. At high pressures, the average energy of the reacting molecules is high, while at low pressures their average energy is close to the threshold energy for reaction. This behaviour is illustrated in Fig. 6.8, in which the product $k(\epsilon)\,n(\epsilon)$ is plotted as a function of energy. The thermal rate coefficients associated with the elementary steps involved in the Lindemann scheme also appear to depend on pressure. The underlying reason for this is that both the population of $A^*(\epsilon)$ and the reaction rate coefficient, $k(\epsilon)$, depend on energy.

The data on which Fig. 6.7 and 6.8 are based are obtained from the RRK model of unimolecular reactions.

The foregoing discussion provides a rationale for one of the main limitations of the energy independent Lindemann mechanism of unimolecular reactions.

The RRK expression for $k_1(\epsilon)$

For the assumptions of RRK theory, see Section 5.3.

To predict the pressure dependence of the unimolecular reaction rate coefficient an expression is required for $k_1(\epsilon)$. This can be obtained from the equation for the populations in the excited levels *in the high pressure limit* (eqn 6.6), when the energized molecule is in thermal equilibrium with the reactants.

The partition function for s identical harmonic oscillators, appearing in the denominator of eqn 6.6, may be written

$$q_A = \left(\frac{1}{1 - e^{-h\nu/k_B T}}\right)^s.$$

For closely spaced energy levels ($h\nu \ll k_B T$), this expression may be replaced by the high temperature, classical approximation for the vibrational partition function,

$$q_A \approx \left(\frac{1}{1 - (1 - h\nu/k_B T)}\right)^s = \left(\frac{k_B T}{h\nu}\right)^s.$$

The density of states employed in eqn 6.6 has been derived previously (see eqn 5.9). Substituting these expressions into eqn 6.6, after a little manipulation, yields

$$\frac{k_1(\epsilon)\,d\epsilon}{k_{-1}(\epsilon)} = \frac{1}{(s-1)!}\left(\frac{\epsilon}{k_B T}\right)^{s-1} e^{-\epsilon/k_B T}\,\frac{d\epsilon}{k_B T}. \tag{6.7}$$

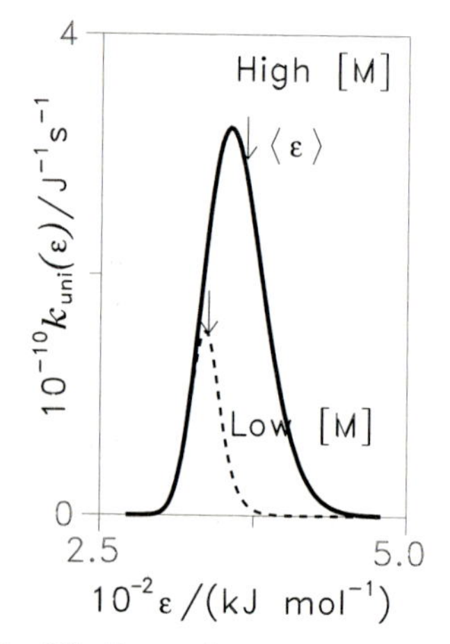

Fig. 6.8 Energy-dependent contributions to k_{uni} as a function of pressure.

RRK theory makes the strong collision assumption for the deactivation rate coefficient, $k_{-1}(\epsilon)$, which is therefore given by the collision number, Z^0_{AM}.

The rate coefficient ratio, eqn 6.7, is equivalent to the population distribution at infinite pressure (see eqn 6.6). The dependence of this distribution on the number of oscillators s is illustrated in Fig. 6.9. As s increases, the distribution is seen to broaden and to peak at higher energies (the maxima, and the energies at which they occur, are given in Table 6.3).

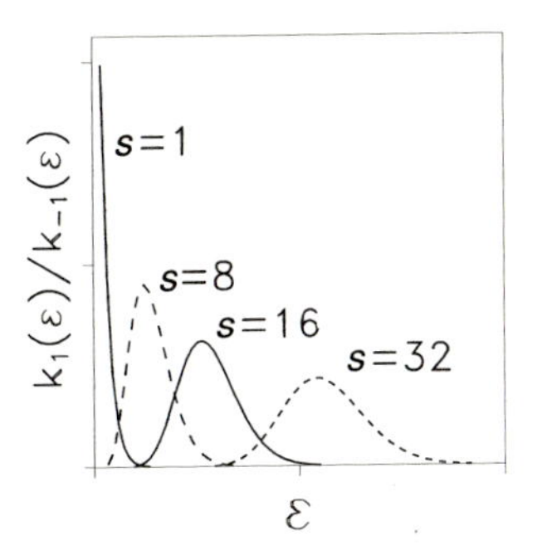

Fig. 6.9 The variations in the populations at infinite pressure with the number of oscillators, s.

The Hinshelwood expression. Integration of eqn 6.7 over energy yields an expression for the average rate coefficient ratio, $k_1(T)/k_{-1}(T)$, *in the high pressure limit,*

$$\frac{k_1(T)}{k_{-1}(T)} = \int_{\epsilon_0}^{\infty} \frac{k_1(\epsilon)\,d\epsilon}{k_{-1}(\epsilon)} \simeq \frac{1}{(s-1)!}\left(\frac{\epsilon_0}{k_B T}\right)^{s-1} e^{-\epsilon_0/k_B T}. \qquad (6.8)$$

The final expression is a reasonable approximation to the integral when $\epsilon_0 \gg s k_B T$. It was first derived by Hinshelwood, although in the Hinshelwood theory of unimolecular reactions, it was assumed valid at all pressures.

This equation gives the fraction of molecules (containing s identical vibrational degrees of freedom) at thermal equilibrium which possess an energy in excess of the reaction threshold, ϵ_0. If one associates $k_{-1}(T)$ with the collision number, eqn 6.8 may be rearranged to give an expression for the average activation rate coefficient in the limit of high pressures. It may be written in the form given by simple collision theory (eqn 1.3), with inclusion of a 'steric factor' defined as

$$P = \frac{1}{(s-1)!}\left(\frac{\epsilon_0}{k_B T}\right)^{s-1}.$$

Table 6.3 The dependence of the maximum in $k_1(\epsilon)/k_{-1}(\epsilon)$ on s

s	ϵ_{max}	$k_1(\epsilon_{max})/k_{-1}(\epsilon_{max})$
1	0	$1/k_B T$
2	$k_B T$	$0.37/k_B T$
3	$2 k_B T$	$0.27/k_B T$
10	$9 k_B T$	$0.13/k_B T$

Values for this factor are given in Table 6.4, for a typical reaction threshold energy of 100 kJ mol^{-1}. As the molecular complexity increases, the probability of finding a molecule above the threshold energy, ϵ_0, increases dramatically, and $P \gg 1$. The activation rate coefficient is thus much *greater* than that predicted by simple collision theory, in contrast with the behaviour observed for bimolecular reactions.

Table 6.4 Hinshelwood 'steric factors', P, for $\epsilon_0 = 100$ kJ mol^{-1} and $T = 298$ K

s	P
1	1
2	40
3	810
10	7.8×10^8

The RRK expression for k$_{uni}$(T). A general, if somewhat cumbersome, expression for $k_{uni}(T)$ is obtained on substituting the expressions 5.10 and 6.7 into eqn 6.4. The RRK theory-predicted pressure dependence of $k_{uni}(T)$ can be illustrated by a *Lindemann plot*, a plot of $1/k_{uni}(T)$ versus 1/[M] (see Fig. 6.10). The experimental data[65] are for the *cis–trans* isomerization of but-2-ene. To obtain agreement with the experimental fall-off curves, s has been treated as an adjustable parameter, and (as seen in Section 5.3) usually takes a value of about half the actual number of oscillators in the molecule. Although RRK theory has been largely superseded by more sophisticated theories, such as RRKM theory, the equations presented above are still used to fit and parameterize experimental data.

The RRK expression for k$_0$(T). In the limit of low pressures ([M] → 0), the unimolecular rate coefficient reduces to the expression

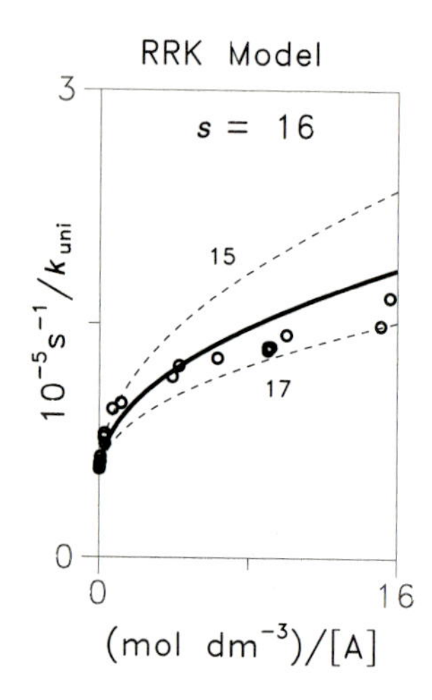

Fig. 6.10 A Lindemann plot of the unimolecular rate coefficient for the *cis–trans* isomerization of but-2-ene.

$$k_0(T) = \int_{\epsilon_0}^{\infty} k_1(\epsilon)\,[\mathrm{M}]\,\mathrm{d}\epsilon$$
$$= Z^0_{\mathrm{AM}}\,[\mathrm{M}]\,\frac{1}{(s-1)!}\left(\frac{\epsilon_0}{k_\mathrm{B}T}\right)^{s-1}\mathrm{e}^{-\epsilon_0/k_\mathrm{B}T},$$

where the strong collision assumption has been employed for $k_{-1}(T)$. The limiting zero-pressure rate coefficient, $k_0(T)$, depends solely on the rate coefficient for the activation step and the concentration of the bath gas, M, and is therefore primarily sensitive to the rovibrational character of the energized molecule, A*; it is independent of the nature of the activated molecule, $A^\ddagger$. Measurement of $k_0(T)$ thus provides a useful means of assessing the validity of the strong collision assumption and of the expressions used for the rovibrational density of states $\rho(\epsilon)$.

The RRK expression for $k_\infty(T)$. In the high pressure limit ([M] → ∞), $k_{\mathrm{uni}}(T)$ reduces approximately to the simple equation

$$k_\infty(T) = \int_{\epsilon_0}^{\infty} \frac{k_1(\epsilon)}{k_{-1}(\epsilon)}\,k(\epsilon)\,\mathrm{d}\epsilon \simeq k^\ddagger\,\mathrm{e}^{-\epsilon_0/k_\mathrm{B}T}. \tag{6.9}$$

$k^\ddagger(\equiv \nu)$ may now be associated with the limiting high pressure *A*-factor, A_∞. Because RRK theory provides no means of calculating $k^\ddagger$ *a priori*, it too must be treated as an adjustable parameter. In practice, A_∞ values vary by orders of magnitude from one reaction to another, as illustrated by the data given in Table 6.5. The inability of RRK theory to provide a rationale for this variation represents a further limitation of the model, and one for which we have to turn to RRKM theory to rectify.

Table 6.5 A_∞-factors for some unimolecular reactions. Also given are the derived ratios of transition state to reactant partition functions assuming a temperature of ~ 500 K

Reaction	A_∞/s^{-1}	$q^\ddagger/q_\mathrm{A}$
$C_3H_8 \rightarrow CH_3 + C_2H_5$	1.5×10^{17}	14400
△ → (cyclopropane → propene, structural formulae)	3.2×10^{15}	310
$(CHO)_2 \rightarrow 2CO + H_2$	1.0×10^{14}	96
ethyl ester (structural formula) → carboxylic acid (OH) + C_2H_4	2.0×10^{11}	0.019

RRKM theory in the high pressure limit

We saw in Section 5.3 that RRKM theory is a *microcanonical* (fixed energy) transition state theory for unimolecular reactions, and the procedures by which $k(\epsilon)$ may be evaluated within this theory have already been discussed. More exact evaluation of the reactant density of states, $\rho(\epsilon)$, than that performed in RRK theory, also allows refined determination of the activation rate coefficient $k_1(\epsilon)$ via eqn 6.6. In the high pressure limit, where A and A* are in equilibrium, RRKM theory yields the following expression for $k_\infty(T)$ (see eqn 5.5):

$$k_\infty(T) = \frac{1}{hq_A} \int_{\epsilon_0}^{\infty} N^\ddagger(\epsilon)\,\mathrm{e}^{-\epsilon/k_\mathrm{B}T}\mathrm{d}\epsilon.$$

Since $N^\ddagger(\epsilon)$ is simply the sum of transition states, this equation reduces to that expected on the basis of CTST, namely

$$k_\infty(T) = \frac{k_\mathrm{B}T}{h}\frac{q^\ddagger}{q_A}\,\mathrm{e}^{-\epsilon_0/k_\mathrm{B}T}. \tag{6.10}$$

This expression should be compared with the RRK expression, eqn 6.9. The pre-exponential factor in the RRKM expression is equivalent to $k^\ddagger$ in RRK theory.

As is the case when CTST is applied to bimolecular reactions, evaluation of the terms in eqn 6.10, in particular $q^\ddagger$, requires knowledge of the transition state moments of inertia and vibrational frequencies. More qualitatively, some insight into the structure of the transition state can be obtained by inverting experimentally determined A-factors to yield ratios of TS to reactant partition functions. Examples are provided in Table 6.5.

Reactions which have large A-factors possess relatively large transition state partition functions. We may interpret the partition function as an estimate of the average number of thermally accessible quantum states at temperature, T. Thus large A-factors correlate with transition states with small energy level spacings and low rovibrational frequencies. Such *loose* transition states are typically found in bond fission reactions, which proceed over surfaces without a barrier to the reverse recombination reaction. The transition states in these cases occur late along the reaction coordinate, when the fragments have already separated to relatively large distances, and TS motions which correlated with reactant bending and rocking vibrations are better thought of as low frequency free, or hindered, rotations.

By contrast, reactions with low A-factors have relatively low TS partition functions. Therefore, they also possess *tight* transition states with comparatively high frequency modes. Such behaviour is characteristic of reactions which proceed via cyclic transition states, since motions which are described as hindered rotations in the reactants become relatively high frequency vibrations in the transition state. These reactions usually proceed over surfaces with large barriers, located early along the reaction coordinate.

Appendices

A.1 Classical elastic scattering of two atoms

The equations of motion

The CM Hamiltonian, eqn 2.6, may be written in the polar coordinates, defined in Fig. A.1. This is achieved by writing $\mathbf{P}^2 = \mu^2(\dot{R}_x^2 + \dot{R}_y^2 + \dot{R}_z^2)$, where $\dot{R}_i$ represents the time derivative of the ith Cartesian component of R, in terms of the time derivatives of R, ϑ and φ, using the following relationships:

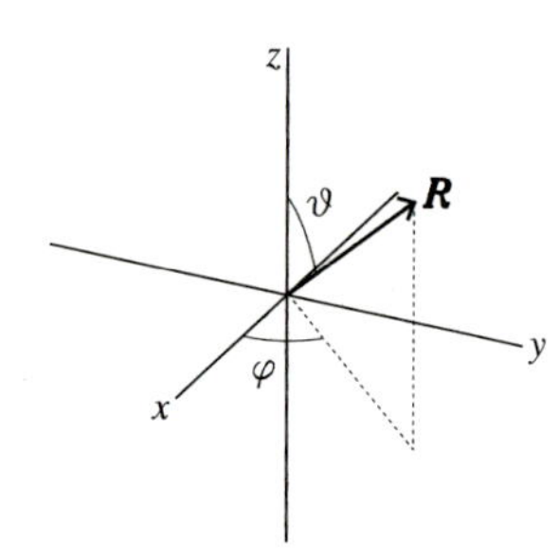

Fig. A.1 Polar coordinates for atom–atom scattering.

$$R_x = R\sin\vartheta\cos\varphi$$
$$R_y = R\sin\vartheta\sin\varphi$$
$$R_z = R\cos\vartheta.$$

The resulting expression is

$$\epsilon_t = \frac{1}{2}\mu\dot{R}^2 + \frac{L^2}{2\mu R^2} + V(R), \tag{A.1}$$

where

$$L^2 = \mu^2 R^4\left(\dot{\vartheta}^2 + \sin^2\vartheta\dot{\varphi}^2\right),$$

and represents the square of the classical orbital angular momentum.

Eqn A.1, in conjunction with the second of Hamilton's equations 2.4, may be used to prove that the orbital angular momentum of the system is conserved. This result can also be derived from the definition of the angular momentum

$$\mathbf{L} = \mathbf{R}\times\mathbf{P} = \mu\left(\mathbf{R}\times\dot{\mathbf{R}}\right),$$

where '×' represents the cross or vector product of the two vectors. If the time derivative of this equation is taken using the chain rule, the result is

$$\dot{\mathbf{L}} = \frac{\mathrm{d}\mathbf{L}}{\mathrm{d}t} = \mu\left[(\dot{\mathbf{R}}\times\dot{\mathbf{R}}) + (\mathbf{R}\times\ddot{\mathbf{R}})\right] = 0.$$

The first term in square brackets is zero because the vector $\dot{\mathbf{R}}$ is necessarily parallel to itself (and hence the cross product must be zero), while the second term is zero because the acceleration, $\ddot{\mathbf{R}}$, which is generated by the potential, is directed along $\mathbf{R}$. Thus both the magnitude of the orbital angular momentum and its direction are conserved (constant) during the collision. The conservation of the direction of $\mathbf{L}$ implies, further, that the relative motion of the particles is confined to a plane, analogous to planetary motion, as noted in Chapter 3. Because of this, we can arbitrarily define the plane such that $\varphi = 0$ and $\dot{\varphi} = 0$, and the orbital angular momentum may be written simply as

$$L = \mu R^2\dot{\vartheta}. \tag{A.2}$$

Because the orbital angular momentum is conserved, the centrifugal kinetic energy, given by the second term in eqn A.1, must increase as the particles approach one another.

The magnitude of the initial angular momentum, L, may be specified in terms of the impact parameter, b, as

$$L = \mu|\mathbf{R} \times \dot{\mathbf{R}}| = \mu R \sin\vartheta\, v_r = \mu v_r b. \tag{A.3}$$

Substitution of eqn A.3 into eqn A.1, and rearranging for $\dot{R}$, yields a differential equation for the radial motion

$$\dot{R} = \frac{dR}{dt} = \pm\left(\frac{2}{\mu}\left[\epsilon_t - \left(\epsilon_t \frac{b^2}{R^2} + V(R)\right)\right]\right)^{1/2}. \tag{A.4}$$

The second differential equation, that for the angular motion, is given simply by rearranging eqn A.2

$$\dot{\vartheta} = \frac{d\vartheta}{dt} = \frac{L}{\mu R^2}. \tag{A.5}$$

The deflection function, χ

Eqns A.4 and A.5 can be integrated numerically for any arbitrary potential energy curve, $V(R)$, to obtain the trajectory of the atoms, for a given set of initial conditions. Alternatively, the two equations may be combined to obtain an equation relating the *deflection angle*, χ, of the scattered atoms to the impact parameter. The deflection angle is related to the angle, ϑ_c, the angular coordinate of $\mathbf{R}$ at the separation R_c of closest approach, via the equation (see Fig. A.2):

$$\chi(\epsilon_t, b) = \pi - 2\vartheta_c. \tag{A.6}$$

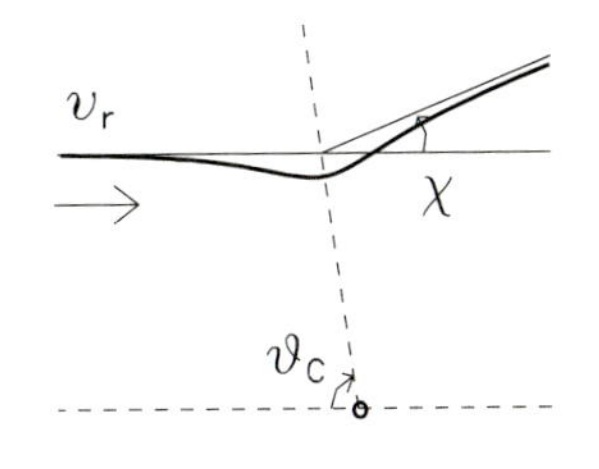

Fig. A.2 The relationship between the deflection angle, χ, and the angle, ϑ_c for atom–atom scattering.

The separation of closest approach may be obtained from eqn A.4 by setting the radial velocity $\dot{R}$ to zero,

$$R_c^2 = b^2\left[1 - \frac{V(R_c)}{\epsilon_t}\right]^{-1}.$$

Finally, ϑ_c, may be obtained by combining eqns A.4 and A.5, using

$$\vartheta = \int \frac{d\vartheta}{dt} dt = \int \frac{d\vartheta}{dt}\frac{dt}{dR} dR,$$

from which we obtain

$$\vartheta_c = b \int_{R_c}^{\infty} \frac{dR}{R^2\left[1 - \frac{b^2}{R^2} - \frac{V(R)}{\epsilon_t}\right]^{1/2}}. \tag{A.7}$$

Eqns A.7 and A.6 reveal the intimate connection between the impact parameter of the collision and the angle through which the departing fragments are deflected. In general, the calculation of the impact parameter dependence of the deflection angle, for all but the simplest of potential energy curves, requires numerical integration of eqn A.7. Such a calculation is illustrated in Fig. 3.4, in which the impact parameter dependence of the scattering angle is shown. The scattering angle is defined as the modulus of the deflection angle, $\theta = |\chi|$. The equation for the classical differential cross-section, eqn 3.7, in conjunction with eqns A.6 and A.7, may be employed to evaluate the differential cross-section, as illustrated in Fig. 3.7.

A.2 Quantum mechanical reaction cross-section

The classical reaction cross-section is expressed as an integral of the opacity function over impact parameter. For example, the initial state i specific excitation function could be written

$$\sigma_i(\epsilon_t) = \int_0^{b_{max}} P_i(b)\, 2\pi b\, db,$$

where $P_i(b)$ is the opacity function for reactants in state i. Quantum mechanically, the initial state specific integral reaction cross-section can be expressed as a sum of (energy dependent) reaction probabilities over the *orbital angular momentum* of the reactants

$$\sigma_i(\epsilon_t) = \frac{\pi}{k_i^2} \sum_0^{\ell_{max}} (2\ell + 1)\, P_i(\ell),$$

where ℓ is the orbital angular momentum quantum number. k_i is the *wavevector* for the relative translational motion of the reactant molecules in initial state i, and is defined in terms of the de Broglie wavelength, λ_i, and the reactant linear momentum, p_i, thus

$$k_i^2 = \frac{4\pi^2}{\lambda_i^2} = \frac{p_i^2}{\hbar^2} = \frac{2\mu\epsilon_t}{\hbar^2}.$$

Note that k_i^2 has dimensions of 1/area.

The quantum mechanical expression for the reaction cross-section reduces to the classical one, if L is assumed to be continuous, rather than quantized (which is particularly appropriate for reactions which have reaction probabilities peaking at high ℓ),

$$L = \hbar\sqrt{\ell(\ell + 1)} \approx \hbar(\ell + \tfrac{1}{2}) = \mu v_r\, b = \hbar k_i\, b,$$

where we have equated the classical linear momentum μv_r with $\hbar k_i$, with the help of the de Broglie equation. In the classical limit, we may replace the sum over orbital angular momentum quantum numbers by an integral:

$$\sigma_i(\epsilon_t) \sim \frac{\pi}{k_i^2} \int_0^{\ell_{max}} (2\ell + 1)\, P_i(\ell)\, d\ell \equiv \frac{\pi}{k_i^2} \int_0^{b_{max}} P_i(b)\, 2k_i^2 b\, db,$$

which reduces to the classical equation for the reaction cross-section given above.

For a reaction (such as A + BC) the total angular momentum, rather than the orbital angular momentum, is conserved. For a given reactant orbital angular momentum quantum number, ℓ, and BC rotational angular quantum number, j, the total angular momentum quantum number, J, must lie in the range

$$|\ell - j| \le J \le \ell + j.$$

In order to conserve total angular momentum, an analogous expression also relates the product momenta, with quantum numbers ℓ' and j', to J.

A.3 Cumulative reaction probability

In Chapter 1, the thermal rate coefficient was expressed in terms of the state-to-state reaction cross sections, by means of a complicated series of summations over initial and final states, and an integral over collision energy. By changing the order of the summations and integration, and transforming the integral to one in total energy, it is possible to rewrite the rate coefficient in the desired form; i.e.

The necessary transformations are performed explicitly in Levine and Bernstein's *Molecular reaction dynamics and chemical reactivity.*

$$k(T) = \frac{1}{hq_r} \int_0^\infty N(\epsilon)\, e^{-\epsilon/k_B T}\, d\epsilon.$$

The resulting relationship between $N(\epsilon)$ and the reaction cross-section can be written

$$N(\epsilon) = h \sum_i g_i\, \rho(\epsilon_t) \sum_f v_r\, \sigma_{if}(\epsilon_t),$$

where g_i is the degeneracy of initial state i (which, for example, might be the $2j + 1$ degeneracy appropriate for an unoriented diatomic molecular reactant with rotational quantum number, j). The sums in this equation are over all energetically accessible reactant and product (f) states which have

$$\epsilon = \epsilon_t + \epsilon_i = \epsilon_t' + \epsilon_f.$$

In the expression for $N(\epsilon)$, $\rho(\epsilon_t)$ represents the translation *density of states* per unit volume. It is defined such that $\rho(\epsilon_t)\, d\epsilon_t$ is the number of translational states in the energy range $\epsilon_t \to \epsilon_t + d\epsilon_t$ (see eqn 4.5),

The derivation of this expression for the density of translational states per unit volume is based on the quantum mechanical solution to a particle confined to a three-dimensional box.

$$\rho(\epsilon_t)\, d\epsilon_t = \frac{4\pi\mu(2\mu\epsilon_t)^{1/2}}{h^3}\, d\epsilon_t\, .$$

From the definition of the translational density of states, the factor $h\,\rho(\epsilon_t)v_r$ may be rewritten more compactly as k_i^2/π, and thus $N(\epsilon)$ reduces to

$$N(\epsilon) = \sum_i g_i \frac{k_i^2}{\pi} \sigma_i(\epsilon_t),$$

where

$$\sigma_i(\epsilon_t) = \sum_f \sigma_{if}(\epsilon_t).$$

Substitution for the quantum mechanical expression for the initial state specific reaction cross-section (given in the preceding section) allows the cumulative reaction probability to be written

$$N(\epsilon) = \sum_i \sum_\ell g_i\, (2\ell + 1)\, P_i(\ell) = \sum_n g_n\, P_n(\epsilon),$$

where the sum over n runs over all reactant states, *including* orbital angular momentum states. This equation provides further justification for the interpretation of $N(\epsilon)$ as a cumulative reaction probability. It also emphasizes that, fundamentally, it is a quantum mechanical entity, although the above equations do provide a means by which it may be calculated classically, via the state-to-state reaction cross-sections.

Bibliography

Background reading

General texts

Andrews, D. L. (1992). *Lasers in chemistry*, 2nd edn. Springer, Berlin.

Atkins, P. W. (1998). *Physical chemistry*, 6th edn. Oxford University Press.

Atkins, P. W. and Friedman, R. S. (1997). *Molecular quantum mechanics*, 3rd edn. Oxford University Press.

Herzberg, G. (1966). *Electronic spectra and electronic structure of polyatomic molecules*. Van Nostrand, New York.

Hollas, J. M. (1982). *High resolution spectroscopy*. Butterworth, London.

Hollas, J. M. (1991). *Modern spectroscopy*. Wiley, New York.

Richards, W. G. and Scott, P. R. (1994). *Energy levels in atoms and molecules*, Oxford Chemistry Primer Series. Oxford University Press.

Softley, T. P. (1994). *Atomic spectra*, Oxford Chemistry Primer Series. Oxford University Press.

Trevena, D. H. (1993). *Statistical mechanics*, Ellis Horwood, Chichester.

Wayne, C. E. and Wayne, R. P. (1996). *Photochemistry*, Oxford Chemistry Primer Series. Oxford University Press.

Wayne R. P. (1988). *Principles and applications of photochemistry*. Oxford University Press.

Wayne R. P. (1991). *Chemistry of atmospheres*. Clarendon Press, Oxford.

Reaction kinetics and dynamics texts

Benson, S. W. (1976). *Thermochemical kinetics*. Wiley, New York.

Bernstein, R. B. (1982). *Chemical dynamics via molecular beam and laser techniques*. Clarendon Press, Oxford.

Beynon, J. H. and Gilbert, J. R. (1984). *Applications of transition state theory to unimolecular reactions*. John Wiley, Chichester.

Billing, G. D. and Mikkelsen K. V. (1996). *Molecular dynamics and chemical kinetics*. John Wiley, New York.

Gilbert, R. G. and Smith, S. C. (1990). *Theory of unimolecular and recombination reactions*. Blackwell, Oxford.

Laidler, K. J. (1987). *Chemical kinetics*, 3rd edn. Harper and Row, New York.

Levine, R. D. and Bernstein, R. B. (1987). *Molecular reaction dynamics and chemical reactivity*. Clarendon Press, Oxford.

Pilling M. J. and Seakins P. W. (1995). *Reaction kinetics*. Oxford University Press.

Pilling M. J. and Smith I. W. M. (1987). *Modern gas kinetics*. Blackwell, Oxford.

Robinson, P. J. and Holbrook, K. A. (1972). *Unimolecular reactions*. John Wiley, London.

Smith, I. W. M. (1980). *Kinetics and dynamics of elementary gas reactions*. Butterworth, London.

Steinfeld, J. I., Francisco, J. S., and Hase, W. L. (1989). *Chemical kinetics and dynamics*. Prentice Hall, Englewood Cliffs.

References

1. Miller, W. H. (1993). *Accounts of Chemical Research* **26**, 174.
2. Alagia, M., Balucani, N., Casavecchia, P., Stranges, D., Volpi, G. G., Clary, D. C., Kliesch, A., and Werner, H.- J. (1996). *Chemical Physics* **207**, 389.
3. Liu, B. (1973). *Journal of Chemical Physics* **58**, 1924; Siegbahn, P. and Liu, B. (1978). *ibid.* **68**, 2457; Truhlar, D. G. and Horowitz, C. J. (1978). *ibid.* **68**, 2466.
4. Stark, K. and Werner, H.-J. (1996). *Journal of Chemical Physics* **104**, 6515.
5. Ho, T.-S., Hollebeek, T., Rabitz, H., Harding, L. B., and Schatz, G. C. (1996). *Journal of Chemical Physics* **105**, 10472; Walch, S. P. and Harding, L. B. (1988). *Journal of Chemical Physics* **88**, 7653.
6. Manolopoulos, D. E. (1997). *Journal of Chemical Society Faraday Transactions.* **93**, 673.
7. See 'Structure and dynamics of reactive transition states' (1991). *Faraday Discussions of the Chemical Society*, **91**.
8. Soep, B. Whitham, C. J., Keller, A., and Visticot J. P. (1991). *Faraday Discussions of the Chemical Society*, **91**, 191.
9. Weaver, A., Metz, R. B., Bradforth, S. E., and Neumark, D. M., (1990). *Journal of Chemical Physics* **83**, 5352; Manolopoulos, D. E., Stark, K., Werner, H.-J., Arnold, D. W., Bradforth, S. E., and Neumark, D. M. (1993). *Science* **262**, 1852.
10. Michael, J. V. and Fisher, J. R. (1990). *Journal of Physical Chemistry* **94**, 3318.
11. Mielke, S. L., Lynch, G. C., Truhlar, D. G., and Schwenke, D. W., (1994). *Journal of Physical Chemistry*, **98**, 8000.
12. Aoiz, F. J., Bañares, L., Díez-Rojo, T., Herrero, V. J., and Sáez Rábanos, V. (1996). *Journal of Physical Chemistry* **100**, 4071.
13. Karplus, M., Porter, R. N., and Sharma, R. D. (1965). *Journal of Chemical Physics* **43**, 3259.
14. Murphy, E. J., Brophy, J. H., Arnold, G. S., Dimpfl, W. L. and Kinsey, J. L. (1979). *Journal of Chemical Physics* **70**, 5910.
15. Mestagh, J. M., Visticot, J. P., and Suits, A. G. (1995). In *The chemical dynamics and kinetics of small radicals* (ed. K. Liu and A. Wagner), Part II, p. 668, World Scientific, Singapore.
16. Brouard, M. and Simons, J. P. (1995). *The chemical dynamics and kinetics of small radicals* (ed. K. Liu and A. Wagner), Part II, p. 795. World Scientific, Singapore; Orr-Ewing, A. J. and Zare, R. N., *ibid.* Part II, p. 936.
17. Neumark, D. M., Wodtke, A. M., Robinson, G. N., Hayden, C. C., and Lee, Y. T. (1985). *Journal of Chemical Physics* **82**, 3045.
18. Faubel, M., Martíncz-Haya, B., Rusin, L. Y., Tappe, U., Toennies, J. P., Aoiz, F. J., and Bañares, L. (1996). *Chemical Physics* **207**, 227; *ibid.* 245.
19. Aoiz, F. J., Bañares, L., Herrero, V. J., Sáez Rábanos, V., Stark, K., and Werner, H.-J. (1994). *Chemical Physics Letters* **223**, 215.
20. Castillo, J. F., Manolopoulos, D. E., Stark, K., and Werner, H.-J., (1996). *Journal of Chemical Physics* **104**, 6531.
21. Gillen, K. T., Rulis, A. M., and Bernstein, R. B. (1971). *Journal of Chemical Physics* **54**, 2831.
22. Rulis, A. M. and Bernstein, R. B. (1972). *Journal of Chemical Physics* **57**, 5497.

23. McDonald, J. D., LeBreton, P. R., Lee, Y. T., and Herschbach, D. R. (1972). *Journal of Chemical Physics* **56**, 769.
24. Schnieder, L., Seekamp-Rahn, K., Borkowski, J., Wrede, E., Welge, K. H., Aoiz, F. J., Bañares, L., D'Mello, M. J., Herrero, V. J., Saez Rábanos, V., and Wyatt, R. E. (1995). *Science* **269**, 207.
25. Parish, D. D. and Herschbach, D. R. (1973). *Journal of American Chemical Society* **95**, 6133.
26. Alexander, A. J., Aoiz, F. J., Brouard, M., Burak, I., Fujimura, Y., Short, J., and Simons, J. P. (1996). *Chemical Physics Letters* **262**, 589.
27. Che, D.-C. and Liu, K. (1995). *Journal of Chemical Physics* **103**, 5164.
28. 'Orientation and polarization effects in reactive processes' (1989). *Journal of Chemical Society Faraday Transactions 2* **85,** 925; (1993). *Journal of Chemical Society Faraday Transactions* **89**, 1401.
29. Parker, D. H., Chatravorty, K. K., and Bernstein, R. B., (1981). *Journal of Physical Chemistry* **85**, 466.
30. Kudla, K. and Schatz, G. C. (1995). In *The chemical dynamics and kinetics of small radicals* (ed. K. Liu and A. Wagner), Part I, p. 438. World Scientific, Singapore.
31. Ionov, S. I., Brucker, G. A., Jaques, C., Valachovic, L., and Wittig, C. (1992). *Journal of Chemical Physics* **97**, 9486.
32. Smith, I. W. M. (1982). *Journal of Chemical Education* **59**, 9.
33. Polanyi, J. C. (1972). *Accounts of Chemical Research* **5**, 161; Polanyi, J. C., and Wong, W.H. (1969). *Journal of Chemical Physics* **51**, 1439.
34. Odiorne, T. J., Brooks, P. R., and Kaspar, J. V. (1971). *Journal of Chemical Physics* **55**, 1980.
35. Hancock, G., Ridley, B. A., and Smith, I. W. M. (1972). *Journal of Chemical Society Faraday Transactions 2* **68**, 2117.
36. Polanyi, J. C., and Woodall, K. B. (1972). *Journal of Chemical Physics* **57**, 1574.
37. Maylotte, D. H., Polanyi, J. C., and Woodall, K. B. (1972). *Journal of Chemical Physics* **57**, 1547; Parr, C. A., Polanyi, J. C., and Wong, W. H. (1973). *Journal of Chemical Physics* **58**, 5.
38. Polanyi, J. C., Schreiber, J. L., and Sloan, J. J. (1975). *Chemical Physics* **9**, 403; Pattengill, M. D., Polanyi, J. C. and Schreiber, J. L. (1976). *Journal of Chemical Faraday Transactions 2* **72**, 897.
39. Tsekoura, A. A., Leach, C. A., Kalogerakis, K. S., and Zare, R. N., (1992). *Journal of Chemical Physics* **97**, 7220.
40. Elsom, I. R. and Gordon, R. G. (1982). *Journal of Chemical Physics* **76**, 3009.
41. Moore, C. B. and Smith, I. W. M. (1996). *Journal of Physical Chemistry* **100**, 12848.
42. Adelman, D. E., Shafer, N. E., Kliner, D. A. V., and Zare, R. N., (1992). *Journal of Chemical Physics* **97**, 7323.
43. Zhang, R., van der Zande, W. J., Bronikowski, M. J., and Zare, R. N. (1991). *Journal of Chemical Physics* **94**, 2704.
44. Crim, F. F. (1996). *Journal of Physical Chemistry* **100**, 12725.
45. Light, G. C. and Matsumoto, J. H. (1978). *Chemical Physics Letters* **58**, 578; Zellner, R. and Steinert, W. (1981). *ibid.* **81**, 568; Glass, G. P. and Chaturvedi, B. K. (1981). *Journal of Chemical Physics* **75**, 2749.

46. Butler, J. E., Jurish, G. M., Watson, I. A., and Wiesenfeld, J. R., (1986). *Journal of Chemical Physics* **84**, 5365.
47. Manolopoulos, D. E. and Light, J. C. (1993). *Chemical Physics Letters* **216**, 18.
48. Seideman, T. and Miller, W. H. (1992). *Journal of Chemical Physics* **97**, 2499.
49. Rabinovitch, B. S., Kubin, R. F., and Harrington, R. E. (1963). *Journal of Chemical Physics* **38**, 405; Placzek, D. W., Rabinovitch, B. S., and Dorer, F. H. (1966). *Journal of Chemical Physics* **44**, 279.
50. Oref, I., Schuelzle, D., and Rabinovitch, B. S. (1971). *Journal of Chemical Physics* **54**, 575.
51. Rynbrandt, J. D. and Rabinovitch, B. S. (1971). *Journal of Physical Chemistry* **74**, 4175; (1971). *Journal of Chemical Physics* **54**, 2275.
52. 'Mode selectivity in unimolecular reactions' (1989). *Chemical Physics* **139**, 1; 'Overtone spectroscopy and dynamics' (1995). *ibid.* **190**, 157; 'Unimolecular reaction dynamics' (1995). *Faraday Discussions of the Chemical Society* **102**.
53. Potter, E.D., Gruebele, M., Khundlar, L. R., and Zewail, A. H. (1989). *Chemical Physics Letters* **164**, 463.
54. Kim, S. K., Lovejoy, E. R., and Moore, C. B. (1995). *Journal of Chemical Physics* **102**, 3202.
55. Jasinski, J. M., Frisoli, J. K., and Moore, C. B. (1983). *Journal of Chemical Physics* **79**, 1312.
56. Michael, J. V. (1990). *Journal of Chemical Physics* **92**, 3394.
57. Park, T. J. and Light, J. C. (1992). *Journal of Chemical Physics* **96**, 8853; Michael, J. V., Fisher, J. R., Bowman, J. M., and Sum, Q. (1990). *Science* **249**, 269.
58. Aoiz, F. J., Bañares, L., Herrero, V. J., and Sáez Rábanos, V. (1997). *Journal of Physical Chemistry* **101**, 6165.
59. The data are reviewed in 'Evaluated kinetic data for combustion modelling' (1992). *Journal of Physical Chemistry Ref. Data* **21**, 411. For low temperature measurements see Frost, M. J., Sharkey, P., and Smith, I. W. M. (1993). *Journal of Physical Chemistry* **97**, 12254.
60. Troe, J. (1996). *Journal of Chemical Physics* **105**, 983.
61. Schatz, G. C., Fitzcharles, M. S., and Harding, L. B. (1987). *Faraday Discuss. Chemical Society* **84**, 359; (1991). *Journal of Chemical Physics* **95**, 1635.
62. Ravishankara, A. R., Nicovich, J. M., Thompson, R. L., and Tully F. P., (1981). *Journal of Physical Chemistry* **85**, 2498.
63. Manthe, U., Seideman, T., and Miller, W. H. (1994). *Journal of Chemical Physics* **101**, 4759.
64. Walch, S. P. and Dunning Jr, T. H. (1980). *Journal of Chemical Physics* **72**, 1303; Schatz, G. C. and Elgersma, H. (1980). *Chemical Physics Letters* **73**, 21.
65. Rabinovitch B. S. and Michel, K. W. (1959). *Journal of American Chemistry Society* **81**, 5065.

Index

Bold numbers refer to figures, italic to tables and 'n' refers to a marginal note.

activation energy 2, 65–66
angular distribution 18–19
angular momentum
  conservation 19, 41–44
  orbital 19, 45, 72–73, 74
anion photoelectron detachment spectroscopy 14–15
Arrhenius equation 2
  see also rate coefficient

Ba + HI 43–44
  see also kinematic constraints
Boltzmann law 6
Born–Oppenheimer approximation 8–9
  breakdown of 13–14

canonical transition state theory
  see transition state theory
capture models 35–37
centre-of-mass frame 16–17, 18–19
  transformation to LAB frame 27, **27**
centrifugal barrier 19–20, 35–36
chemical activation 53–54
classical mechanics 15–17, 19–22, 23–26, 72–73
  see also Hamilton's equations
classical trajectory method 15–21, 23–26
cone-of-acceptance 32
coordinates
  bondlength **11**
  elastic scattering **16**, 72
  Jacobi **11**, 23, **25**, 25*n*
  mass-scaled 40
  reactive scattering 11
correlation rules 13
cross-section
  collision 2
  differential 5, 18–32
  reaction 3–5
  state-to-state 4–5, 33–45
cumulative reaction probability 7, 46–49, 74–75

D + H_2 17, 44, 64
  see also H + H_2
D + I_2 28*n*, 29, 42–43
deflection function 20, 73
densities of states 52, 55–57
dividing surface
  see transition state

effective potential 19, **19**, 35
elastic scattering 15–17, 18–22, 72–73
energy disposal 37, 38–39
energy transfer 2, 5*n*, 34, 52, 66–67, 69–70
  see also elastic scattering
  see also inelastic scattering
energy utilization 33–37, 38
excitation functions 33–37

F + H_2 12–13, 14–15, **27**, 28–29, 41, 49
force 11, 15–17
  see also classical mechanics
force constant 10–11
free energy flow 53–55
  see also intramolecular vibrational redistribution

H + CO_2 32, 65
H + D_2 29–30, 35, 62–63, 64
  see also H + H_2
H + H_2 11-12, 49, 50–51
  see also H + D_2
  see also D + H_2
H + H_2O 10, 17, 45, 66
Hamilton's equations 15–17
  atom–atom scattering 19, 72–73
  reactive scattering 23–26
  see also classical mechanics
harpoon reactions 13, 29, 36–37
Hinshelwood theory 69, 69*n*
hot atom reactions 28, 33

impact parameter 4, 19, 32
inelastic scattering 18
  see also energy transfer
intramolecular vibrational redistribution (IVR) 38, 45*n*, 55
  see also free energy flow
isotope effects 64–65

Jacobi coordinates
  see coordinates
jet cooling 14
  see also molecular beam experiments

K + CH_3I 29, 36–38, **37**
K + I_2 12–13, 29, 36–38, **37**
$k(\epsilon)$
  see rate coefficient
kinematic constraints 28*n*, 39–44
kinetic energy
  collision 2*n*, 19, 34–37
  centrifugal 19
  radial 19

laboratory (LAB) frame 18
  see also centre-of-mass frame
  see also Newton diagram
laser based experiments 27–28, 33–34
Lindemann scheme 66–67
Lindemann plot 69–70, 70
line-of-centres model 34–35
long-lived complexes 30–31

Maxwell–Boltzmann distribution 6, 46
minimum energy path 12
  see also reaction coordinate
molecular beam experiments 18, 26–27, 33
molecular orbital theory 9–10

$N(\epsilon)$
see cumulative reaction probability
$N^{\ddagger}(\epsilon)$
see transition state theory
Newton diagram 27, **27**
see also centre-of-mass frame
Newton's laws
see Hamilton's equations
see classical mechanics
non-adiabaticity 13–14

O (^{1}D) + H_2 13, 31, 45
O (^{3}P) + HCl 44–45
OH + H_2
see H + H_2O
OH + CO
see H + CO_2
opacity function 4, 43–44
opacity model 22

partition functions 6, 7, 48, 60, *62*, 70–71
phase space theory 38–39
photoelectron spectroscopy
see anion photoelectron detachment spectroscopy
Polanyi's rules 37–38
populations
thermal 6–7
see also statistical models
see also cross-sections
potential energy
diatomics 10–11
Eckart 47*n*
harmonic 10–11
Lennard-Jones 11
parabolic 46–47
surfaces 8–17
triatomics 11–13
prior distribution 38–39

quantum dynamical calculations 17, 74–75
elastic scattering **21**, 22

rainbow scattering 20–22
rate coeffient
canonical 60–71
microcanonical 52–59
pressure dependence 66–70
temperature dependence 2, 65–66, 69–70
thermal 2, 5–7, 60–71
Rb + CH_3I 32
reaction coordinate **11**, 12
see also transition state
reaction probability 4, 7
see cumulative reaction probability
reactions
association 1–2, 53–54
bimolecular 1, 17–52, 60–66, 72–75
elementary 1
light attacking/departing atom 42–44
light atom transfer 42
unimolecular 1, 52–59, 66–71
with barrier 33–34
without barrier 34–37
recrossing trajectories 49, 51–52
relaxation
see energy transfer
resonance scattering 15, 22, 48–49
RRK theory 54–59, 66–70
RRKM theory 54–55, 66–68, 70–71

scattering angle 5, 18–19
versus impact parameter 20, 73
simple collision theory 2, *3*, 34–35
comparison with transition state theory 60–61
single collision conditions
see energy transfer
skew angle 40–41, 45
spectroscopy based experiments
see laser based experiments
statistical models
energy disposal 38–39
see also RRK and RRKM theories
see also transition state theory
stereochemistry 31–32
steric factor 3, *3*, 61–62
stripping reaction 44*n*
strong collision assumption 67
sum of states 49–50, 55–57
surprisal analysis 39

transition state (TS) 49
spectroscopy 14–15
theory (TST) 49–52, 60–66, 70–71
see also RRK and RRKM theories
tunnelling 47, 64
resonant 22, 48–49

van der Waals complexes 11, 14, 32

weak collisions 67